FUNDAMENTAL RESEARCH IN
ULTRA HIGH DILUTION AND HOMOEOPATHY

Fundamental Research in Ultra High Dilution and Homoeopathy

edited by

Jurgen Schulte
Department of Applied Physics,
University of Technology,
Sydney, Australia

and

P. Christian Endler
Ludwig Boltzmann Institute for Homoeopathy,
Graz, Austria

KLUWER ACADEMIC PUBLISHERS
DORDRECHT / BOSTON / LONDON

A C.I.P. Catalogue record for this book is available from the Library of Congress.

ISBN 0-7923-5051-0

Published by Kluwer Academic Publishers,
P.O. Box 17, 3300 AA Dordrecht, The Netherlands.

Sold and distributed in North, Central and South America
by Kluwer Academic Publishers,
101 Philip Drive, Norwell, MA 02061, U.S.A.

In all other countries, sold and distributed
by Kluwer Academic Publishers,
P.O. Box 322, 3300 AH Dordrecht, The Netherlands.

Printed on acid-free paper

Printed in the Netherlands

To my mother.

Jurgen Schulte

This work has been supported in parts by

The Pro-Vice Chancellor Discretional Fund,
University of Technology, Sydney,

and

The Bundesministerium für
Wissenschaft und Verkehr, Vienna.

Jurgen Schulte, Dr Sci, is Director of Physical Sciences at the Ludwig Boltzmann Research Site for Low Energy Bioinformation in Graz (Austria), Lecturer in Applied Physics at the University of Technology, Sydney (UTS), Member of the UTS Centre for Materials Technology and Member of the UTS Centre of Biomedical Technology as well as consultant to the European Union research subcommittee HOMEOMED.

P. Christian Endler, Dr Phil, is Director of Biological Sciences and of Administration at the Ludwig Boltzmann Research Site for Low Energy Bioinformation in Graz (Austria), Research Associate at the Ludwig Boltzmann Institute for Homoeopathy in Graz, and Associate Professor at the Institute for Holistic Medicine at the University of Urbino (Italy). He is Member of the International Group for High Dilution Research (GIRI), as well as consultant to a European Union working group for the program BIOMED.

CONTENTS

PREFATORY WORD

THE INFORMATION MEDICINE HYPOTHESIS

Peter Fisher

Of all the so-called 'complementary' or 'alternative' therapies, homoeopathy poses by far the greatest challenge to current scientific understanding. At least, it challenges scientific understanding if you believe there is anything particular to understand. Although many doctors concede that patients often feel that they have been helped by homoeopathy, the prevailing consensus in the medical and scientific community has long been that any improvements attributed by patients to homoeopathy are, in fact, entirely due to 'placebo' or non-specific effects. In other words homoeopathic medicines have no real effect, the results experienced by patients are entirely due to expectation, suggestion, encouragement, psychotherapy, plain good advice or other factors associated with the process of homoeopathic treatment.

However, the evidence of recent placebo-controlled clinical research makes it quite clear that this is not an adequate explanation. The metanalysis of clinical trials of homoeopathy by Linde et al, which forms the final contribution to this volume, is of high quality, and concludes unequivocally that the effects of homoeopathy cannot be attributed solely to placebo effects. Two previous independent surveys of the clinical research literature have come to similar, positive conclusions. Yet homoeopathy remains highly controversial, and has been accused of violating scientific laws, including the most fundamental, such as the Second Law of Thermodynamics, it is claimed that acceptance of homoeopathy would entail 'rewriting the textbooks'. Why exactly is it that the claims made for homoeopathy provoke such scepticism? In order to understand this, we need to examine certain assumptions which underlie the current biomedical paradigm and which appear to conflict with homoeopathy.

Homoeopathy is a pharmaceutical therapy in the sense that it involves the administration, usually by mouth, of medicines. For most doctors, pharmacologists and other medical scientists, drugs are necessarily chemical. The idea of medicines acting in any other way than by some form of chemical reaction has never arisen for most doctors and medical scientists and appears bizarre, if only because it is so unfamiliar. Current pharmacological concepts are dominated by the so-called 'lock-and-key' model of drug action, in which a drug molecule (the key) interacts with a receptor molecule (the lock), 'turning', or, more commonly, 'jamming' it. Other models of drug action, such as differential toxicity (as with antibiotics and cytotoxic drugs) can be identified, but all depend on the chemical action of drug molecules.

The core of the controversy surrounding homoeopathy is its use of very high dilutions. Homoeopathic medicines are prepared by a process, sometimes known as dynamisation or potentization, which involves serial dilution with succussion. Succussion is vigorous shaking, with striking against an elastic surface (traditionally a leather-bound book). Although the substances to be diluted and the diluents are precisely specified in homoeopathic pharmacopoeias, surprisingly, there is no standardisation of the rate or amplitude of succussion. The dilutions are usually prepared in steps of 1:10 (denoted x or dH) or 1:100 (c or cH). Dilutions of 30cH and higher are in common use by homoeopaths.

Anyone with a basic knowledge of chemistry will appreciate that this is a problem for those for whom medicines are chemicals: a 30c dilution is 10^{-60} of the starting material. Avogadro's Constant, the number of particles (atoms or molecules) in a gram mole of any substance, is of the order of 10^{-23}. The implication is clear, and not disputed by homoeopaths: any dilution above 23x or 12c is very unlikely to contain even a single molecule of the starting substance, whose name appears on the label. Such dilutions are termed ultramolecular. In practice the ultramolecular limit is passed at much lower dilutions for many substances (because of initial concentration etc). In any case the sources of many homoeopathic medicines are complex mixtures of molecules of variable molecular weight and concentration. Even for pure substances of known molecular weight and concentration the 'ultramolecular limit' can only be stated in terms of probability.

Clearly, the mechanism of action of homoeopathy cannot be a key-and-lock interaction, since there is no 'key'. To those accustomed to thinking of the action of medicines in exclusively chemical terms, this appears to be an excellent reason why homoeopathy cannot possibly have any real effect. So deep-rooted and unquestioned is the assumption that medicines must act by chemical means that the tone of criticism of homoeopathy has sometimes gone well beyond expressions of scepticism, and has instead been couched in emotive terms, inappropriate to scientific discourse.

A more rational strand of criticism is based on Bayes Theorem, whereby the probability of a proposition being true can be calculated from its 'prior' probability and the strength of the evidence, to give a 'posterior' probability. In other words, implausible propositions require strong proof in order to be accepted. Of course, 'plausibility' may be somewhat arbitrary, and vary between individuals. In the case of homoeopathy one could frequently substitute 'unfamiliar' for 'implausible'. Nevertheless scepticism concerning homoeopathy is greatly amplified by the lack of a generally accepted theoretical framework to account for the effects observed in placebo-controlled clinical trials. From the Bayesian perspective, there are two ways in which the scientific impasse around homoeopathy might be resolved: either more and stronger empirical evidence; or a plausible theoretical framework, which would increase the prior probability.

In the last few years, opinion within the homoeopathic scientific community has converged around the 'Information Medicine hypothesis', as the most promising theoretical approach to the mode of action of homoeopathic medicine. The Information Medicine hypothesis can be simply stated as: 'Water, (and perhaps other polar solvents), can under certain conditions, retain information about substances with which they have previously been in contact, and subsequently transmit this information to presensitized biosystems.' In this conception, the medicine is physical rather than chemical in nature, and its action is understood in terms of cybernetic processes, instead of key and lock interactions. This accounts for the basic principle of homoeopathy, the treatment of 'like with like' (similia similibus curentur), in terms of regulatory feedback, as well as the problem of ultramolecular dilutions.

The Information Medicine hypothesis has two essential components: the storage of information by water, and its reception and processing by biological systems. If the former has been little investigated, the latter is, if anything, even less explored! Although information processing by biosystems has been extensively investigated at the neurological and genetic levels, relatively little work has been done at the inter- and intracellular level, and again it is generally assumed that communication at this level is exclusively chemical. *Fundamental Research in Ultra High Dilution and Homoeopathy* brings together a series of contributions which examine these two aspects of the Information Medicine hypothesis, their interfaces and implications.

The popularity of homoeopathy is growing worldwide, and its potential is increasingly recognised, but for too long it has languished in a scientific limbo, for lack of a coherent, plausible theoretical underpinning. In this and their previous book[1], Jurgen Schulte and Christian Endler have assembled a group of high-level authors of diverse backgrounds who are prepared to confront the challenging scientific problems posed by homoeopathy. This is an exciting book about an exciting area of scientific endeavour, it marks an important step in the opening up of a whole new domain of biomedical knowledge. I strongly recommend *Fundamental Research in Ultra High Dilution and Homoeopathy* to anybody interested in grasping current theoretical and basic research aspects of this fast-evolving area of science.

Peter Fisher
Director of Research
Royal London Homoeopathic Hospital
Great Ormond Street
London WC1N 3HR
England

Fax +44 171 833 7212
Email pfisher@gn.apc.org

[1] Endler P and Schulte J eds., Ultra High Dilution: Physiology and Physics. Kluwer, Dordrecht 1994.

PREFACE

Jurgen Schulte and Christian Endler met in 1990 at an international conference on the *Structure of Water* held in the Lecture Halls of the University of Graz (Austria). Disappointed by the lack of a systematic strategy of research into the physics of homoeopathy Jurgen Schulte started to work on the establishment of scientifically acceptable research standards in physics of homoeopathy and encouraged academic researchers to establish a coordinated and focused research strategy. In 1994, with the help of major representatives of the international research community, they edited one of the first academic interdisciplinary books on Ultra High Dilution and homoeopathy that underwent a rigorous scientific international referee process before publishing. Due to the dedicated help of the prominent referees (BD Josephson, Nobel Laureate, Cavendish Lab., Cambridge; M Bastide, Fac de Pharmacy, University Montpellier; RG Jahn, Aerospace Science, Princeton University), the book 1994 was quickly considered a mile stone and turning point for the scientific approach of research into Ultra High Dilution and homoeopathy. Since then the academic research community has grown considerably and many international conferences have been held. Today, research into homoeopathy is to be accepted by the European Union as part of the academic sciences, worthy to be funded at European Union level; an effort that took many years of research coordination and research strategy development. Excerpts of the Research Strategy of the European Committee for Homoeopathy (ECH) have been included in this book.

In this volume we present a collection of current efforts addressing the fundamental problem in homoeopathy and Ultra High Dilution research, namely the concept of biologically relevant information transfer and storage in Ultra High Dilutions and physiological systems. As there is a multitude of parameters involved in possible information storage in aqueous solutions, and an equal multitude of approaches to address this problem, we tried to focus only on those models that are based on established academic science and their standard verification processes. In a research field that has received academic interest only recently, it is quite natural that during its initial state the vast amount of problems to be tackled may sometimes result in uncoordinated research, especially when cross-disciplinary research ranging over more than two faculties is involved. Communication problems and lack of conceptual understanding of other science faculties makes it difficult to address a scientific problem that is inherently multidisciplinary.

For that reason, we are addressing some fundamental (theoretical) problems of information transfer and storage in ultra high diluted systems at the beginning of the book, setting the ground for stimulating possible questions to be asked in the following more experimental section. We feel that after more than two centuries of trying to verify effects of homoeopathic remedies and the recent successes of bio-assay experiments, it is of utmost importance to develop a fundamental understanding of the underlying physics responsible for producing the measured results. Thus, in this book we place physical models with relevance to homoeopathy and ultra high dilutions next to each other and, wherever possible, tried to refer to direct links between them. The results of new and greately extended experiments start to present a consistent pattern. Nevertheless, verification of the experiments is still required, and of course, presented the theories presented are still trying to converge towards a common understanding of the underlying physical processes. In this sense, we are presenting current fundamental approaches in theory, experiments and clinical trials, the foundation of future coordinated research efforts. A *Glossary* at the end of the book which covers technical terms frequently used in physics and physiology is intended to bridge some language barriers among our readers from different science faculties.

We hope that this book will help to stimulate further research as well as interdisciplinary discussion.

We wish to thank all those who directed our interest towards the research field and who critically accompanied our work (T. Kenner, M. Haidvogel, G. Kastberger, G. Karmapa, D. Lama); the authors whose work was included; Suyash Prasad for valuable comments to the manuscript and E. Lauppert who helped us survive the time of editorial difficulties; our families and friends, as well as the publisher. We are especially grateful to L. Johnson, Pro-Vice-Chancellor (Research) at University of Technology, Sydney, for the trust in this work and the financial support.

1998

Sydney, Australia — Graz, Austria

Jurgen Schulte — Christian Endler

FUNDAMENTAL RESEARCH IN ULTRA HIGH DILUTION AND HOMOEOPATHY

Summary

Jurgen Schulte and Christian Endler

This volume gives a critical presentation of physical, biophysical and physiological fundamental model theories as well as biological and medical experimental data with indications of possible low energy bioinformation transfer in ultra high dilution and homoeopathy.

Research into regulatory therapy such as homoeopathy follows either of two complementary lines: the *stimulus applied* and the *reaction cybernetics of the living system*. In their previous presentation of multidisciplinary research into ultra high dilution in physics and physiology, the authors approached the subject based on the hypothesis of a certain *stimulus* or *perturbation* through a base substance being the carrier of physiologically relevant information (Endler and Schulte, 1994). The volume presented here elucidates the *perturbation* hypothesis in more rigorous detail whereby the medical and physiology research focus is on the living system's ability to react.

The two introductory chapters by the European Committee for Homoeopathy and by Wiegant et al. outline the history of homoeopathy and the general strategy for research into homoeopathy as it has been formulated by the research community of the European Union.

Fundamental physical models of information storage and information transfer in water and physiological systems are discussed by Schulte. In this context bio-information transfer finds its place in the mesoscopic regime of the physical science, ie. in the region between quantum and continuum physics. The contribution presents a rigorous treatment of physical concepts from which a variety of disciplines may draw from.

J. Schulte and P.C. Endler (eds.),
Fundamental Research in Ultra High Dilutionand Homoeopathy, 3-7.

With the vast amount of research and well established models available today it may already be possible to write a conventional physiology textbook based on energetic interactions (e.g. the energy metabolism of the cell, voltage shifts on cell membranes, etc.). Nevertheless, energetics is only one part of the physiological interplay, the competing other part is entropic interactions. Thus, we felt it important to include a contribution by Ho which elucidates the role of entropy and its relation to information transfer.

A fundamental model on information storage is presented by Del Giudice and Preparata. This addresses the information storage in aqueous solutions and associated liquids the basis of on electro- and magneto-dynamics. The contribution includes a discussion on the stability of electrodynamic domains as well as unpublished material on magnetic polarization.

The phenomena of resonance, interactivity, feedback and chaos control as well as the step to pathological phenomena are touched upon by Kratky, Bellavite and Signorini. These contributions focus on the complexity of weak electromagnetic interactions and put them into the context of dynamical systems. In dynamical systems theory the dynamics defining parameters themselves are the information carriers as well as the means of information transfer. At this point, the link to physiological systems becomes very obvious which gave us a natural transition to the physiological and medical science contributions which follow, introduced by a chapter summarizing the classification of non-clinical fundamental research into homoeopathy. This chapter has been explicitly composed in order to systematise research strategies with regard to the basic concepts of homeopathy. The definitions set forth here bring a semblance of order into a field of research sometimes criticized for its lack of structure and clarity. The contribution includes an extensive list of references and, between research into the stimulus and the reaction, sets the context for the contributions that follow.

Wiegant and van Wijk present fundamental *reaction cybernetics* of the *"similarity law"* of iso- and homoeopathy on the cell biological level. The contribution on biphase reactions of living systems to the same stimulus when applied at different intensities, known in biology as "hormesis effects", was first discussed in our previous book "Ultra High Dilution" (Endler and Schulte 1994). This volume presents a continuation of the discussion of their work, now set into the framework of much improved international research.

In "Ultra High Dilution" Pongratz and Endler had introduced botanical and zoological bio-assays. This work has evolved over time resulting in considerable improvements on the statistics of their data and coherence of their work. This work has evolved resulting in considerable improvements in the statistics of the data and coherence of their work. The previously observed data are confirmed in the results presented here. The conventional and homoeopathic way to prepare concentrated solutions are compared, and the possibility of information storage in water by electronic means is presented. These papers also discuss the difficulties of experimental work in this area.

In her contribution on information and communication in living organisms, Bastide establishes a working hypothesis concerning the mechanism underlying the observed effects of ultra-high diluted endogenous compounds. Here it is proposed that non-molecular information corresponding to an original molecule is conveyed to a recipient organism which deals with this special cue to restore normal immune and endocrine functions. This fundamental research sets new standards for experimental work in this area.

Clinical effects of homoeopathy are discussed in a paper by Linde where clinical trials are critically reviewed with respect to trial procedures, placebo effects and possible sources influencing data and their interpretation.

We appreciate the multitude of backgrounds that our readers are coming from. Therefore we have included a *Glossary* at the end of the book covering frequently used, terms in physics and homoeopathy. Due to the limited space available in such continuing compressed presentation of fundamental research in this area, all contributors are aware of that some aspects of their topics may have been covered in more detail than others. Thus, we would like to encourage readers to contact individual authors using the provided *List of Authors* at the end of the book.

Conclusion and Outlook

This volume includes discussions on a multitude of blind placebo-controlled studies. Considering their international origin the results obtained appear too homogeneous to be purely coincidental. However, some of the more important studies have *failed* to confirm earlier results. For this reason it seems appropriate to speak of *clear indications rather than of proof of the correctness of certain homoeopathic concepts*.

An overall assessment of the accomplished work shows that both in biological and clinical experiments the effects of homoeopathic agents as compared to their matched controls (placebo) are only marginal. If, for example, only studies complying to high methodological standards are considered, then the global difference in effect is around or below 10%.

This overall outcome may speak in favour of the homoeopathic concepts under study, but even in cases where findings are backed up by statistically significant differences, they are nowhere near as convincing as the successes reported by homoeopathy in practice. Various explanations have been offered to account for this discrepancy:

so-called "research results" are purely coincidental;

the successes reported by homoeopaths only refer to a few isolated cases;

fundamental research is irrelevant to homoeopathy;

affirmative results are usually not reproducible, that is, in the best case they are credible isolated incidents; in the worst, they are attributable to inadvertent or even wilful distortion.

In our opinion none of the above arguments finds support in the material we have compiled in this volume. However, a review of the presented results, both non-clinical and clinical, does suggest that research into homoeopathy requires a well structured research strategy, if we are looking for reliable conclusive results.

In physical sciences non-linearity and seemingly small effects are an inherent part of the trade of physics which can be treated in a systematic way on a well founded base of physical science. The advantage of physical sciences over clinical trials is that results are easily reproducible due to the open books of the laws of fundamental physics. This volume shows that the laws of physics as such give room for possible information storage and information transfer in associated liquids. The open questions here are how stable and unique the stored information is, how an information transfer or propagation is being kept pure throughout a physiological system, and which is the physiological useful part of the stored information is, if it can be proved to be stable. Furthermore, if information is transferred from an original chemical substance, can it be synthezised and coded by other means?

Selection of clinical material or patients/test subjects to be examined

Clinical homoeopathic research requires that certain fundamental principles are observed: the principle of *individualisation*, the choice of an appropriate *symptomatology*, and a certain level of *vital resources* still present in the patient. These requisites have been consistently observed in recent clinical studies.

Furthermore, experimental proof of homoeopathic effects presupposes the use of a sufficiently sensitive model. In fundamental research this limits the range of organisms suitable for research (e.g., simple organisms under stress); in research on humans too, it may require a certain degree of sensitivity in the patient or test subject, similar perhaps to that of electromagnetically sensitive patients described in the conventional medical literature.

Drug proving, the actual empirical basis of homoeopathy (see Walach in "Ultra High Dilution"), is probably subject to the same kind of uncertainty. In our view the current practice of testing homoeopathic drugs by provoking proving symptoms in healthy test subjects selected at random should be complimented by the more goal oriented strategy of conducting preliminary screening experiments to select particularly sensitive subjects. Some authors in the older homoeopathic literature endorsed the idea of restricting drug proving to sensitive subjects (e.g., the Materia Medica of T.F. Allen), a notion which certainly deserves thorough investigation.

In conclusion we may confidently say that homoeopathy has begun to subject itself to the criteria of conventional research and that understanding of homoeopathy today has advanced considerably from where it was only a few decades ago. At that time the scientific standards of homoeopathic research were rightly rated as poor. The results we have to offer today may still be far from conclusive, nevertheless a very important goal has been achieved, i.e. research methods and standards have become more reliable and reproducible and a well implemented peer review system is in place.

References

Endler PC, Schulte J, Ultra High Dilution: Physics and Physiology, Kluwer Academic Publishers, Dordrecht (The Netherlands), 1994.

FUNDAMENTAL CONCEPTS AND PRACTICAL APPLICATION OF HOMOEOPATHY

The European Committee for Homoeopathy

A compilation with the help of the members of the ECH

A Short History

Homoeopathy is a system of medical practice that originated with the work of the early nineteenth-century German doctor Samuel Hahnemann. He discovered that diseases can interact in very special ways. They can temporarily or permanently take each others place. An example is the relation between eczema and asthma. It is well known that these two chronic expressions of an allergic constitution can alternate.

Hahnemann studied a lesser well known variety of the interaction between disease, viz. the lasting substitution of one disease by another. He observed, for instance, that a skin eruption that had existed for a long time, vanished after measles. A more recent and almost unknown example is normalisation of essential hypertension for several months after a febrile disease. Hahnemann wondered what made the difference between temporary and permanent substitution of one disease by another, and he discovered that the lasting substitution occurred when the two diseases had similar symptoms. This phenomenon has become known as the 'Law of Similars' (The term 'law' is not yet clearly defined. From an epistemological point of view it is not yet clear when a regularity may be called a law of nature. However, in homoeopathy the term 'law' is used, because the Law of Similars and Hering's Law are used as such).

The next step for Hahnemann was to try to apply this substation in a therapeutic way. As he was also an expert in chemistry, he was familiar with a lot of symptoms caused by toxic agents and he was aware of the fact that certain diseases, as they occur in nature, resemble strongly the symptoms of intoxication. For instance, a belladonna intoxication resembles scarlet fever, a quinine intoxication malaria, and an arsenic intoxication cholera. It was only a small step for Hahnemann to combine

J. Schulte and P.C. Endler (eds.),
Fundamental Research in Ultra High Dilutionand Homoeopathy, 9-18.

the ideas of substitution of similar diseases with substitution by means of an intoxication. For instance, he tried to cure patients with scarlet fever with belladonna. Then his suspicions proved to be correct that diseases could be cured with substances that produce similar toxic effects, a dose of belladonna in scarlet fever, a dose of arsenic in cholera. He assumed that for certain diseases specific remedies could be found. Therefore he searched for other potentially helpful drugs and tested them extensively on volunteers ('drug proving').

Gradually Hahnemann's ideas about homoeopathy became more refined. For instance, he discovered that patients with other diseases than cholera could also be cured by arsenic, provided they showed other common 'arsenic characteristics'. And, on the other hand, that not all cholera patients reacted to arsenic, but needed another remedy, depending on their individual symptoms. So he changed the nosological (i.e., the systematic classification of diseases) frame of thinking with a concept of specific, individual states of illness.

He perceived that patients homoeopathically apparently cured of a certain disease could have relapses of this disease, or even get other diseases. Only if other data as from constitutional and psychological features and previous diseases were also taken into account for the selection of a remedy, a permanent cure could be brought about.

Besides, he found out that diluting remedies in a special way (potentisation) not only reduced or abolished their toxic effect, but also, paradoxically, increased their curative effect.

A contemporary German doctor, Constantin Hering, noticed that during the process of recovery previous diseases could recur in the reverse order of their appearance in a patient's personal life. Recovery from chronic complaints turned out to proceed from within outwards (e.g., in a patient with asthma a previous eczema can (re)appear in the process of recovery and is the last to vanish). This regularity is called 'Hering's Law'.

Bearing in mind this historical evolution of understanding, a definition of homoeopathy can be given.

Definition

Homoeopathy is a system of medical practice aimed at methodologically improving the level of health of an organism by the administration of proven* potentised** medicines, which are individually selected in accordance with the Law of Similars***.

* A proven substance is one that has been pharmacologically tested on healthy human beings (see section pharmacology 3.7).

** Potentised = processed in a specific way, namely by succussion of serial dilutions (see Glossary at the end of this book).

*** Law of Similars = a substance, which is capable of provoking symptoms in a healthy organism, acts as a curative agent in a diseased organism in which the same symptoms are manifested.

The symptoms of a diseased organism, the 'disease picture', are classified and interrelated in such a way as to trace patterns that match with the 'remedy picture' (i.e., the symptoms of a remedy provoked in a healthy organism). The more peculiar and individual the symptoms are, the more indicative of a certain remedy. Common symptoms are the least important in this process of matching 'disease picture' and 'remedy picture'.

If the similarity is great enough, the original disease or complaint does not recur after discontinuing the administration of the remedy. In case of incomplete similarity, only partial or temporary effects are noticeable. Hering's Law (see section *A Short History* and *Consultation*) is the criterion for the assessment of the long-term therapeutic effect.

Hierarchy of Functions

Practical experience has shown that a hierarchy of functions plays a role in the application of the Law of Similars. Symptoms (i.e., perceived by the patient's own senses and conveyed to the doctor during case taking) and signs (i.e., perceived by the doctor's senses and found during the physical examination) that arise from higher functions appear to be more indicative within the frame of reference of the law of Similars. The process of classifying and interrelating symptoms and signs is called hierarchisation.

Symptoms/signs that correspond with higher hierarchical levels are:

- Mental and emotional symptoms (e.g., fears, misconceptions of reality).
- Causative factors (i.e., symptoms that originate from, for example, a reaction to grief, cold, trauma, etc.).
- Detailed specification of symptoms:
    - Modifying factors ('modalities') (i.e., environmental factors that change the symptom (e.g., weather, temperature, noise, etc.).
    - Concomintants (i.e., symptoms associated in time with the main affection).
    - Sensations, localisations, and extensions of pains.

A further category of features that can help to choose a remedy are symptoms or properties of a person's inherited constitution. These are reflections of a person's susceptibility of disease:

- Personal constitutional features (e.g., prone to chilliness, perspiring feet, aversion to milk).
- Modifications of the constitution arising from previous diseases, accidents, vaccinations, effects of lifestyle, etc.).
- Family History

Even if the patient seeks help for somatic complaints (like warts), all hierarchical levels of the patient (mental, emotional, and physical) are methodically screened and the highest level of disturbance (from a hierarchical point of view) is detected. Unspecified local symptoms correspond with the lowest hierarchical level.

Prescribing homoeopathic remedies or remedies of botanical origin is not identical with homoeopathy. Homoeopathy is to be distinguished (e.g., from anthroposphic medicine and phytotherapy). In anthroposophic medicine, homoeopathic remidies are used among others, but in a completely different way and not based on drug provings and the Law of Similars. In phytotherapy, the Law of Similars is not considered either (the plant extracts are used in their crude form and not, as in homoeopathy).

A 'Living' Example

The following example will illustrate the depth and possibilities of homoeopathy. A man or woman comes to see a doctor, having had stomach pain for a considerable period of time. The stomach is very sensitive to pressure. There is a feeling of pressure in the stomach, as of a stone, several hours after eating. The pressure of tight clothes is not well tolerated. Coffee and alcohol, usually liked very much, are not tolerated now. The patient has nausea and a sour taste in the mouth in the morning. She/he cannot sleep, owing to thoughts about work. She/he is very irritable, especially in the morning on waking. She/he is impatient, competitive, ambitious (a 'workaholic'), easily offended, sensitive to noise and angry if objects are not in their proper place. She/he likes to smoke, eat spicy food, and drink alcohol and coffee. In general, she/he is a person prone to chilliness and is easily constipated.

For a homoeopathic doctor, the diagnosis 'peptic ulcer' (seen of the X-ray) is not enough. On the contrary, all the above mentioned information is important and comprises the 'disease picture'. The appropriate homoeopathic prescription for this patient is Nux Vomica. This remedy will not only stimulate healing of the peptic ulcer, but will also, even in the first place, improve the patient's well-being (i.e., the patient will become much more relaxed and will be able to sleep soundly, again. Actually, the objective of a homoeopathic treatment is not only relief of a certain complaint, but amelioration of the patient's health in general.

Concept of Health and Disease

The example in the previous section indicates that in homoeopathy the physical pathological changes at the material level, in this case a peptic ulcer, is only a small part of the information needed to apply the Law of Similars. Also, a great many other signs and symptoms appeared to be indispensable. The peptic ulcer belongs to the material aspect of disease and can be seen in a dead body that only consists of matter. The other symptoms mentioned can only exist in an individual that is alive and cannot be explained by the mere existence of the peptic ulcer alone. Most of these symptoms had even existed before a peptic ulcer developed. Thus, homoeopathic prescribing is based on the symptoms belonging to the dynamic aspect of disease, rather than the material aspect.

Generally speaking, the first symptom of disease (= dis-ease) is a feeling of uneasiness. Later on, disease starts revealing itself through a succession of symptoms. As the disharmony progresses, more definite indications start appearing, first in the form of subjective symptoms (i.e., abnormal sensations and complaints), then followed by objective symptoms ('signs') (i.e., abnormal data, like disordered functioning of affected organs and systems, and later on even structural alterations in the tissues and organs, detectable by means of tests and other measuring procedures). Subjective and objective symptoms together furnish the 'external' evidence of disease. In homoeopathy, these very symptoms point to a curative remedy.

Homoeopathy assumes the concept that functional disturbances always precede structural changes in tissues and organs. Thus, pathological changes do not represent the cause of disease, but its effects.

The basis in homoeopthic thought is that health is not a static condition, but a stage of development (i.e., a dynamic process that tends to maintain a state of optimum equilibrium). This concept presupposes a built-in self-regulation mechanism that protects against loss of balance. Disease reflects an intensified attempt to restore an out-of-balance state, resulting from disturbing physical, chemical, biological, and emotional factors. Disease is conditioned by susceptibility and it manifests itself through symptoms in the mental/intellectual, emotional, and physical plane. The self-regulation mechanism is regarded as responsible for protection against the loss of balance, as well as for its restoration. All that a doctor can do towards a cure is to the process of restoration, by stimulating this self-regulation mechanism.

Homoeopathic remedies appear to induce a process of reorganisation of vital functions, by stimulating this self-regulation mechanism. This reorganisation can result in a complete cure in cases where only functional derangement had caused the symptoms. The more structural the changes that have been caused by the disregulation, the more partial a recovery will be. If a disease process has come to an end and the tissue damage has become irreversible, homoeopathy will only have a palliative effect.

A homoeopathic treatment can be rapidly successful in acute illness, but in chronic illness full recovery may take a considerable time in view of the necessary internal reorganisation of the patient's constitution.

Indications for Homoeopathic Treatment

As homoeopathy stimulates the self-regulation mechanism, it is easy to see that , in principle, all illnesses can benefit from homoeopathic treatment. The therapeutic possibilities, of course, depend on the extent to which the organism is able to recover. Limitations are specific surgical indications, deficiency diseases, and very serious diseases, in which gross anatomical changes, like a tumour, have evolved. However, in incurable stages of a disease homoeopathic treatment may relieve a patient to a large extent.

The experience of homoeopathic doctors is that usually many diseases, ranging from asthma to arthritis, pneumonia, migraine, goitre, liver diseases, depression, etc., can be successfully treated with homoeopathic remedies. Homoeopathy offers therapeutic options where conventional treatments have failed or plateaued, where they do not exits for the problem, or where they are contraindicated or not tolerated.

Homoeopathic treatment is compatible with other medication, but a homoeopathic doctor seeks to reduce medication to a minimum. Moreover, homoeopathy can often be used as a first option in certain problems, keeping more costly and potentially toxic treatment as a second option.

Consultation

The homoeopathic consultation consists of a general medical part and a specific homoeopathic part.

General Medical Part of the Consultation

Homoeopathic doctors make a conventinal diagnosis like all other doctors,. conventional diagnosis helps to choose the case for homoeopathic treatment, and to determine the location and the pathological changes. It also picks out the characteristic symptomatology athat denotes the total picture of the state of illness. The knowledge of the type and the extent of the pathological changes, as well as the knowledge of the natural curse of diseases guides the doctor's prognostication. Besides this, it renders inestimable help inplanning the homoeopathic treatment. It gives an idea of the location 9tissues and organs affected), the pathological changes (tyupe, degree, and extent), the pathogenic agent, the physiological disturbances induced, the psychological accompaniments, and the patho-genesis. It is obvious that no rational treatment can be planned in the abaence of knowledge of th efactors mentioned above. Similarly, ikt will be impossible to give the prognosis with any semblance of accurate, unlesss the doctor possesses this knowledge. It also ensables the doctors to 'choose their case' and to eleminate cases that require primarily surgical or other measures. It also makes assessment of the results possible.

Specific Homoeopathic Part of the Consultation

For this part of the sonsultation, much more time, as well as specific knowledge and skills, are requred. Homoeopathic expertise consists of the combination of:

- The skill in finding the relevant symptoms and constitutional features of a patient ('disease picture') by case taking and case analysis.
- The knowledge of the ;remedy pictures' (see above) and the methodology of finding them, if not recognised at onced.

The homoeopathic part of the consultation can be divided into three phases:

A Case-Taking

The aim of homoeopathic case-taking is to gather all data that are homoeopathically relevant. The most important sources of relevant data are the patient's spontaneous narration of her/his history and an accurate, unprejudiced observation. An unprejudiced observation is one of the most difficult things to achieve. Unconscious expectations and attitudes may bias and colour the doctor's observation and receptivity to a large extent. The different attitudes held by the observing doctor should not interfere with accurate observation. The recording of the observations should also be untinged by interpretation.

It is clear that a homoeopathic doctor requires tact and a mastery of the technique of case taking, in order to obtain all this detailed information about the patient as an individual. It should be appreciated that causative factors in the emotional sphere are all important and no successful homoeopathic prescription can ever be achieved without giving due consideration to them. Thus, a homoeopathic doctor should have a good insight into human nature and a good working knowledge of psychosomatic interrelationships in human illness.

The personal history of the patient enables the doctor to:

- Understand the patient as a person and the way he has become ill.
- Determine the constitutional and familial tendencies that probably operate to the detriment of the patient.

On the physical level, the homoeopathic doctor is not satisfied with conventional diagnosis alone. More detailed information is needed, like the cause of the symptoms, modalities of a symptom, concomitants, constitutional features, family history, previous operations, vaccinations, etc., as explained the above sections.

B Case Analysis and Data Processing

A complete case analysis offers a full comprehension of the patient's personality features, his constitution, and the mechanism of the production of symptoms (both subjective and objective).

As explained in section 3.2, the collected data is classified ('hierarchised') and a corresponding known 'remedy picture' is looked for. In this stage, the repertory plays an important role. The repertory is a reference book (or computer data base) in which correlations between symptoms and remedies are listed in a systematic way.

C Remedy Selection

The accuracy of homoeopathic therapy consists in finding the remedy that matches best with the highest level of disturbance in the organism (see above). The insistence on the exact similarity leads to the use of the single remedy, as only one remedy can be the most appropriate at a given time. However, in extenuating circumstances where the homoeopathic doctor is unable to obtain the history of all symptoms, or, because of the doctor's level of expertise, she/he is unable to find the correct remedy in time, giving more than one remedy is justified.

Besides the choice of the remedy, the dose and the frequency of administration of the remedy has to be determined.

Follow-Up Consultations

In the follow-up consultations, the patient's response to the remedy has to be accurately assessed. The doctor has to determine the necessity for a repetition of the remedy or for a change of remedy or potency.

The quality of long term case management consists in successively curing the disturbances in the organism, starting from the hierarchically highest disturbed level. A complete recovery usually follow certain patterns, viz. the recovery proceeds from within outwards, from the most important organs to the least important organs. The symptoms recur in the reverse order of their appearance in the patient's life ('Hering's Law).

A STRATEGY FOR RESEARCH INTO HOMOEOPATHY

Fred Wiegant, Dick Koster, Ton Nicolai

This article has been originally drafted by F Wiegant, D Koster and Ton Nicolai as a report of the *European Committee for Homœopathy* to the *Directorate General XII of the European Commission* in Bruxelles. Since the article contains important contributions regarding assessments and standards in research into homœopathy, it has been found that it will be useful to a larger international audience. Hence, we made part of this information available here in the context of the fundamental research presented in this book. For the original report (Wiegant et al, 1997) the European Committee for Homœopathy has asked all individuals that are known to be involved in homœopathic research in the European Union for their additions, comments and suggestions. The final report and this version of it, therefore, can be considered to be the shared view of all homœopathic researchers in the European Union.

Contributions to this article have been made by Madeleine Bastide, Vera Baumans, Martin Dicke, P Christian Endler, Peter Fisher, Robbert van Haselen, Gerard Jansen, K. Keller, Klaus Linde, Dick Koster, Hans Miltenburg, Philippe Servais, Jurgen Schulte, Cyril Smith, David Spence, Aslak Steinsbekk, Jeremy Swayne, Harald Walach, Michel van Wassenhoven, Fred Wiegant, Roel van Wijk and Harry van der Zee.

Homœopathy is claimed to be effective by millions of patients and thousands of homœopathic doctors all over the world. Still, the effectiveness of homœopathy is debated and its mechanism of action not well understood.

The main objective of this article is to provide a strategy for assessing the value of homœopathy for health care in general, and particularly in Europe where it seems be more established than in other regions of the world.

J. Schulte and P.C. Endler (eds.),
Fundamental Research in Ultra High Dilutionand Homoeopathy, 19-43.

The strategy delineated by this article focuses on increasing

- the knowledge of homœopathy
- the understanding of its working mechanism
- the effectiveness of homœopathy

This article has been presented its formal report form to the European Commission because the contents of this report may fit into the fifth European Research framework programme (Biomed III) that is soon to be developed (1997).

In its discussion paper "Inventing Tomorrow - Europe's research at the service of its people" the Commission describes as the aim of forthcoming research to make it "more efficient and increasingly directed towards meeting basic and economic needs by bringing about the changes which each individual citizen desires" (p. 2). "Scientific and technical progress should pave the way for the opening up of new areas, whether they relate to knowledge, ideas, products, processes or services, in order to improve the quality of life for individuals" (p. 6).

The fifth framework programme ("Research at the service of the people") states that one of the basic principles for selecting areas where Community research can play a decisive role, should be "the concepts of *public and social acceptability*, which ensure that research is meaningful to European citizens" (p. 8). In addition, chosen topics should support Community policies, in particular through "expanding knowledge and developing new techniques for *health protection*"(p. 9). "Experience shows that we should be moving from research aimed purely at technological achievement to research aimed at satisfying consumers by providing high quality goods and services which are produced in an acceptable manner at low cost and which are at the same time highly diversified and personalised and rapidly available" (p. 12). One of the priorities of the fifth framework programme is "the targeting of activities and the impact the research will have on people's lives" (p. 13). The far-sighted position that the European Commission has taken on in this regard reflects a general trend that can been seen world wide, however, not as close to its implementation as in Europe.

Without any doubt, research in homœopathy is meaningful to European citizens. In fact, consumer surveys show increasing popularity of non-conventional medicine and especially homœopathy. Both users and critics agree that serious side effects do not occur and that these medicines are safe to take. Because of this safety Council Directive 92/72/EEC allows a simplified registration procedure for most homœopathic medicines.

Besides, homœopathic care is inexpensive and therefore claims to provide substantial savings to health service and welfare budgets. Homœopathy's aim to mobilize the self-healing properties of the body in a curative and/or preventive sense is welcomed by European citizens for their personal health care. In addition, if homœopathy is introduced in the livestock farming sector, the European citizen could also be more protected from pharmacological residues in animal products.

By and large, homœopathy may provide a substantial contribution to improved quality of life for European citizens and thus seems to qualify for obtaining research funding by the European Union.

The objective of this article is to provide a strategy for assessing the value of homœopathy for health care in Europe.

Homœopathy is becoming increasingly popular in the European Union. Millions of consultations are taking place each year. Both users and critics agree that serious side effects do not occur and that these medicines are safe to take. Homœopathic care is inexpensive and therefore claims to provide substantial savings to health service and welfare budgets. Homœopathy's aim to mobilize the self-healing mechanisms of the body in a curative and/or preventive way is welcomed by European citizens for their personal health care. The majority of general practitioners in different European countries feel that homœopathy, like some other forms of complementary medicine, deserves a place in mainstream medical practice. Clearly, in the current climate homœopathy is expected to figure more largely in both primary and secondary health care.

Although encouraging results have been obtained in rigorously performed clinical trials supporting the efficacy of homœopathy, as was shown in recent meta-analyses, homœopathy is still much disputed by mainstream medicine. In order to be accepted as a valid part of medical practice, it is recognized by both opponents and supporters, that more research investment is needed. Research in homœopathy has frequently been hampered by methodological problems as well as by gross underinvestment in the academic resources required.

The Homœopathic Medicine Research Group (a joint group of researchers in mainstream medicine and homœopathy formed by *the Directorate General XII of the European Commission*) recently concluded that homœopathy is certainly researchable. They advised the stimulation of further research in homœopathy and produced *guidelines for review of clinical research protocols* in homœopathy and concluded that highest priority should be given to clinical research in which mainly large trials of sound and rigorous methodology should be done. In addition, that in *fundamental research* previously used model systems in which an effect of high potencies have been claimed, should be repeated.

One of the main objectives of this report is not only to prove homœopathy by effectiveness studies, but also to improve it by trying to understand and evaluate its therapeutic success (or failure) in daily practice. Apart from efficacy, important questions concerning safety, mechanism of action, the value of different therapeutic strategies and costs have to be investigated in a systematic way.

Clinical research

To assess the effectiveness and efficacy of homœopathy, the ECH takes the view that a step-by-step approach is necessary, starting from registration and data collection in daily practice and retrospective research aimed at collection of clinical cases through open outcome studies to randomised clinical trials. The main advantage of this approach is the systematic way in which knowledge is gathered from the population under study, the relevant and promising diagnoses and the early feedback preventing failures based on premature assumptions.

A programme for the post-marketing surveillance of safety and the evaluation of cost-effectiveness, patient satisfaction and/or preference and quality of life should be started.

Testing of homœopathic remedies in so-called homœopathic pathogenetic trials or provings needs methodological improvement. One of the main objectives for research in this field is to establish a "gold" standard. In addition, new remedies are to be tested and currently used remedies should be retested.

The development and introduction where possible of homœopathy in the livestock farming sector would be particularly important with a view to the more effective protection of consumers from pharmacological residues in animal products and better conditions for livestock currently being reared, in view of the rules soon to be introduced concerning organic livestock farming techniques.

Fundamental research

Constraints to the acceptance of homœopathy which are most often cited by opponents are the lack of a scientific basis and the absence of a theoretical model for homœopathy. If one considers the fundamental studies within homœopathy, the large majority (90%) of research is concentrated on demonstrating an effect of potentised agents or remedies. Hardly any research is directed to the similia principle, nor are activities structured in a systematic programme to unravel the underlying mechanism required to explain the effect of high potencies. Therefore, research activities should be intensified in the two fundamental aspects of homœopathy, i.e. the similia principle (remedies or conditions that cause symptoms when applied to healthy biological systems, can be used to treat the same symptoms in diseased biological systems) and the specific way in which remedies are produced (by a process of successive dilution and succussion, which is called potentisation).

For research on high potencies two main activities are required:

- the repetition of successful claims in different model systems
- further elaboration of the information theory and the physical aspects involved in potentisation in order to improve our understanding of how potencies may maintain their information, how this information may interact with the organism and in what way it might be transmitted to exert its curative action.

Current position of homœopathy

Homoeopathy is widely used by citizens of Europe: 32% of the French, 31% of the Dutch and 16% of the British public have reported taking homoeopathic medicines in recent public opinion surveys and its popularity is increasing. In the United Kingdom the market for homoeopathy is growing by 15% annually, growth is more rapid in some southern countries, while in countries with well developed markets (such as France and Germany), the growth rate is slower. The total European market for homoeopathic medicines in 1995 was 1,344 million ECU, the largest markets are Germany and France, with 36 and 33%, respectively of the total market. The European Union has taken steps to create a single European market for homoeopathic medicinal products: EC Council Directives 92/73/EEC and 92/74/EEC applying to homoeopathic medicinal products for human and veterinary use came into force in January 1994.

Implementation of these directives has been patchy among member states, partly because of scientific problems.

Over the past few decades encouraging results have been obtained in well performed clinical trials supporting the efficacy of homœopathy, as was shown in recent meta-analyses (Kleijnen et al. 1991, Linde et al., HMRG report, 1997, short version, p.16-17). Still, homœopathy is much disputed by mainstream medicine.

European Commission Homœopathic Medicine Research Group

In December 1993 the *European Parliament* requested the *European Commission* to conduct a scientific investigation of homoeopathy. *Directorate-General XII* (Science, Research and Development) of the *European Commission* supported an expert group, the Homoeopathic Medicine Research Group (HMRG) to conduct this investigation, with the mandate to discuss methodologies, to undertake a constraint analysis, to develop common standards, and to consider whether homoeopathy is a researchable question.

In their report the HRMG concluded that homœopathy is certainly researchable. They advised the stimulation of further research in homœopathy and produced guidelines for review of clinical research protocols in homœopathy. An overview of clinical research in homoeopathy identified 184 controlled clinical trials of homoeopathy. The statistical meta-analysis of pooled results gave highly significant results in favour of homoeopathy.

The HMRG also examined the "constraints" to the acceptance of homoeopathy by sending a questionnaire to a sample of 1500 decision makers in Austria, Belgium, France, Germany and Great Britain. Despite wide divergence of attitudes to homoeopathy, there was a consensus that more research is required: 77% of supporters and 62% of opponents of homoeopathy favoured more scientific investigation.

Two main fields of research in homœopathy can be identified, namely *clinical research* and *fundamental research*. Clinical research is given highest priority by the HMRG with an emphasis on the randomised clinical trial and on improving methodology in order to establish the efficacy of homœopathy. Their report provides some useful guidelines on methodology for clinical research in homœopathy "with the objective to allow commonly accepted judgments of clinical efficacy, effectiveness and benefit of homœopathy by future clinical research". With respect to fundamental research, the HMRG recommends the replication of previously used model systems in which an effect of high potencies has been claimed.

Objectives of the homœopathic research community

The authors of this report agree on the recommendations of the HRMG, but take the view that a more extensive approach is necessary. Although it is recognized that randomised clinical trials do form an essential part of clinical research activities, hardly any attention so far has been given to registration studies in clinical practice.

In fundamental research the ECH favours an extension of research activities from mere replication activities to the study of the two fundamental aspects of homœopathy (the similia-principle and the characteristic way in which remedies are prepared, both issues being addressed in this volume).

This attitude within the ECH results from an increasing emphasis to open up the so-called "black box" in clinical and fundamental research. Until recently most research in homœopathy has occurred according to this "black box" model. In this model the researcher is mainly interested in the results of treatment, since the only relevant information specified in this type of study concerns the change in condition of patients before and after treatment, without consideration of the mechanisms behind these results or the possibilities to optimize the response. An evident disadvantage of the "black box" approach is the lack of information required to explain failure of clinical trials or fundamental studies. The changing attitude from *proving* to *improving* homœopathy requires the opening-up of the "black box", characterizing the different fields of research and approaching these different aspects of homœopathy in a systematic way. The task of research should not only be to study the effectiveness of a certain intervention, but also to understand the strengths and weaknesses of the diagnostic process, the different steps leading to the selection of a remedy and the various treatment protocols aimed at the promotion of human well being. For this the conditions under which successes and failures occur should be studied closely, which clearly requires careful registration of clinical practice.

Opening-up of the "black box" will increase our knowledge of homœopathy, increase our understanding of the basic mechanism of action of homœopathic remedies and may improve the therapeutic effect by optimizing the clinical methodology. This widened approach in research on the homœopathic method will not only be of benefit for homœopathy, but may also have potential importance to the evolution of medical science as a whole.

Aims and contents of this report

Up till now there has not been, among homœopathic researchers, a common strategy or structured framework, following a step-by-step procedure to evaluate homœopathy. This type of approach to research provides a structure or a strategic "backbone" connecting different fields of research, identifying successive steps which are required to further our understanding of various aspects of homœopathy. The advantage of such an approach is to identify blank spots at an early stage and to improve coherence between research groups or research activities.

The aim of the ECH report therefore was to identify different research activities in clinical and fundamental research to improve, among researchers, a systematic strategy to evaluate different aspects of homœopathy.

This article begins with a brief description of homœopathy as a method and its specific concept of health and disease. This is necessary to understand its conceptual depth and therapeutic possibilities. Then the two main fields of research in homœopathy are described, i.e. *clinical research* (on human beings and animals) and *fundamental research* (in laboratories), followed by the definition of several target areas of homœopathic research.

HOMŒOPATHY AS A METHOD

Basic principles

Homœopathy is a system of medical practice based on two pillars: the similia principle and the use of potentised medicines or remedies.

• *The similia principle*

This principle implies that substances / remedies that cause symptoms when applied to healthy biological systems, can be used to treat the same symptoms in diseased biological systems. The similia principle, which forms the fundamental basis of homœopathy, is unconventional in the sense that it has not been recognized, accepted or studied by mainstream medicine.

• *The use of potentised medicines*

Homœopathic medicines, some of which in their crude state are potentially toxic, are prepared according to a specific process of successive dilutions and succussions (for liquid preparations) and successive triturations (for solid preparations) – this process is called potentisation – in such a way as to nullify their toxic properties. A number of variations exist on both the dilution steps and the way in which succussion is performed.

Specific concept of health and disease

The basis of homœopathic thought is that health is not a static condition but a dynamic process that tends to maintain a state of optimum equilibrium. This concept presupposes a built-in self-regulation mechanism, which protects against a loss of balance. Disease reflects an intensified attempt to restore an out-of-balance state, resulting from disturbing physical, chemical, biological and emotional factors. Disease is conditioned by susceptibility and it manifests itself through symptoms in the mental/intellectual, emotional and physical plane. The self-regulation mechanism is regarded as responsible for protection against the loss of balance as well as for its restoration. All that a doctor can do is to assist the process of restoration, to stimulate this self-regulation or self-recovery mechanism.

Homœopathic medicines are supposed to induce a process of reorganisation of vital functions by stimulating this self-regulation mechanism. This reorganisation can result in complete cure in cases where only functional derangement had caused the symptoms. The therapeutic possibilities, of course, depend on the extent to which the organism is able to recover. The more structural the changes that have been caused by the deregulation, the more partial a recovery will be. Limitations are specific surgical indications, deficiency diseases and very serious diseases in which gross anatomical changes have evolved. If a disease process has come to an end and the tissue damage has become irreversible, homœopathy may only have a palliative or relieving effect.

Homoeopathy is used to treat a wide range of disorders, a 1991 review of the research literature found the following main areas had been scientifically investigated: trauma and pain, respiratory infections, mental and psychological disorders, pollinosis and rheumatology. Homœopathy can even offer therapeutic options where conventional treatments have failed or plateaued, where they do not exist for the problem, or where they are contraindicated or not tolerated.

Homœopathic medicines

Homœopathic medicines, mostly called remedies, are of botanical, chemical, mineral, zoological or microbiological origin. They are prepared from products, substances or compositions, called homœopathic stocks, in accordance with a homœopathic manufacturing procedure described by the pharmacopoeias currently used, officially, in the Member States. They are obtained from stocks by the process of potentisation, i.e. successive dilutions and succussions for liquid preparations and successive triturations for solid preparations.

Homœopathy in clinical practice

Homœopathic treatment is aimed at methodologically improving the level of health of an organism by the administration of proven potentised medicines, which are individually selected in accordance with the similia principle. The symptoms of a diseased organism, the "disease picture", are classified and interrelated in such a way as to trace patterns that match with the "remedy picture", i.e. the symptoms provoked by a remedy in a healthy organism. The more peculiar and individual the symptoms are, the more indicative for a certain remedy. If the similarity is great enough, the original disease or complaint does not recur after discontinuing the administration of the remedy. In case of incomplete similarity only partial or temporary effects are noticeable. Homœopathic treatment is compatible with other medication, but a homœopathic doctor seeks to reduce medication to a minimum.

Practical experience has shown that a hierarchy of functions plays a role in the application of the similia principle. Symptoms and signs that arise from higher functions (like emotional symptoms or individual patterns of responding to environmental factors) appear to be more indicative within the frame of reference of the similia principle. Even if the patient seeks help for somatic complaints (like headache or gastritis), all hierarchical levels of the patient –mental, emotional and physical– are methodically screened and the highest level of disturbance –from a hierarchical point of view– is detected. Both human beings and animals may successfully respond to a homœopathic treatment. However, since mental, emotional and individual symptoms are more difficult to elicit in animals, in veterinary practice symptoms lower in hierarchy often lead to the most appropriate remedy.

CLINICAL RESEARCH

In general, research consists of a stepwise, hierarchical approach from gathering uncontrolled, observational knowledge to controlled, causal studies. In their approach to demonstrate the value of homœopathic therapy and to increase the knowledge of homœopathy, homœopathic researchers wish to concentrate on a number of research fields extending from retrospective and descriptive studies to randomised clinical trials. This approach emphasises the complementary role between effectiveness[1] and efficacy[2] research. These different forms are illustrated in Figure 1 (see next page).

The goal of this approach is to identify areas where homœopathy offers particularly effective treatments to the patients.

The following steps can be distinguished:

- daily registration in homœopathic practice
- selecting possibly good diagnoses and/or remedies (successful groups of patients)
- testing these findings in a series of research
    - from retrospective to prospective
    - from open to randomised controlled studies (including time-series designs)
    - from the initiation of new trials to the replication of well-performed clinical trials of sound methodology.

In addition to the various steps required to design relevant clinical trials, research on different aspects of homœopathy should be further stimulated, such as:

- assessing the value for money of homœopathy, in other words economic evaluation of homœopathy
- describing the remedy picture (homœopathic pathogenetic trial)
- analysis of different aspects of the patient-doctor interaction, and
- increasing our knowledge of various phenomena described within homœopathy such as symptom aggravation, syndrome shift, etc ('auxiliary' studies).

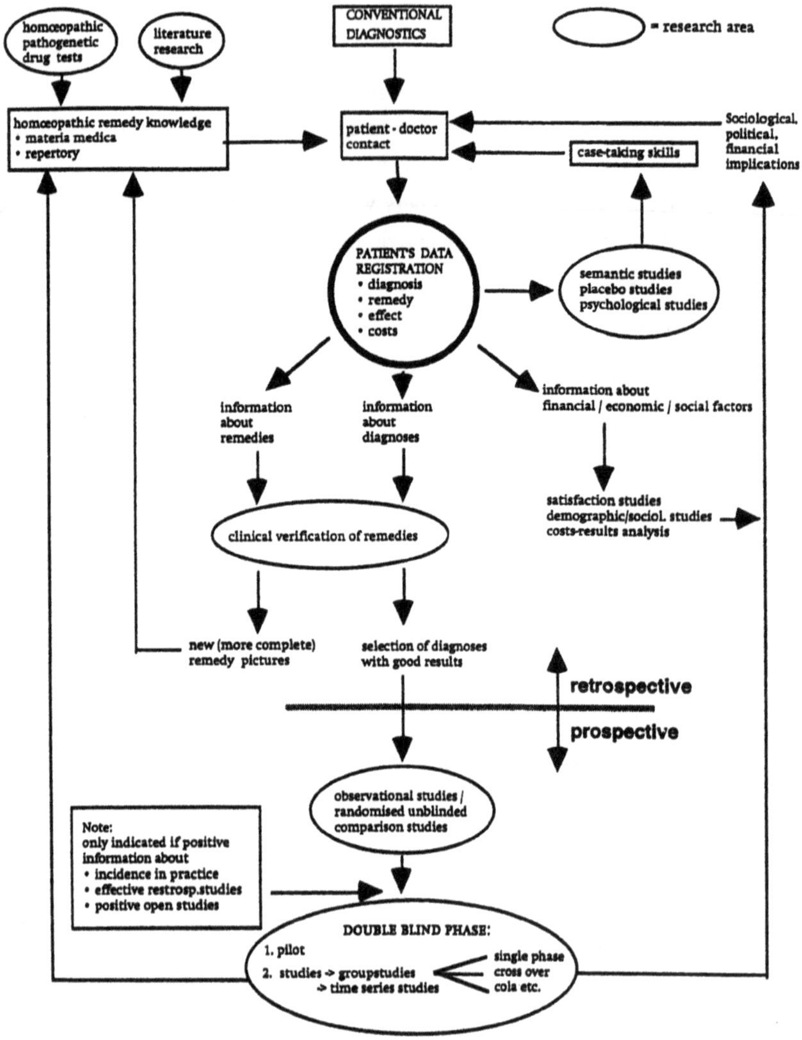

Figure 1. A strategy for clinical research (Koster, 1997)

Cycles of confirmation. In this figure three cycles of confirmation can be distinguished. The main axis (central, vertical) is the proposed step-by-step procedure for clinical research from registration to randomised controlled trials. The left cycle shows the clinical confirmation of our knowledge of homœopathic remedies, leading to a more complete knowledge of them. The right part shows the cycle for financial, economic or social factors. All three cycles have their feedback on the patient-doctor contacts, improving future outcomes.

Effectiveness and efficacy of homœopathy

A relative lack of descriptive observational studies is observed leading to situations where randomised clinical trials are undertaken without proper knowledge of the dynamics within homœopathic practice of the type of patients under study such as: (a) the length and frequency of treatment, b) the dynamics of disease manifestations of patients during treatment, (c) the success of treatment in the specific type of disease under study, etc. Therefore, the importance of observational studies is emphasised by stimulating both prospective and retrospective data collection. These types of studies are relevant to build up a database of experience thus widening the basis for selecting specific types of diseases to be included in clinical trials or in more rigorous experimental and observational studies.

As a first step the ECH recommends daily registration in homœopathic practice and retrospective research aimed at collection of clinical cases. These epidemiological studies are required incorporating a large number of different patients from a large number of homœopaths showing different diagnoses and or diseases followed over a long period of time.

Also, in accordance with the recent call for post-marketing surveillance of conventional drugs, post-prescription clinical verification is becoming an interesting means of assessing the effectiveness as well as the safety of homœopathic therapy.

Although these types of studies are descriptive in nature, they will give insight into the relative effectiveness for different types of disease, etc. rather than giving clues to the cause of therapeutic effects. Moreover, these types of studies will provide the necessary foundation for a further investigation of the cost-effectiveness of homœopathy.

Economic evaluation of homœopathy

Cost-effectiveness is an integral part of health care policy influencing both medical and administrative decisions. There is some evidence available that homœopathy may offer good "value for money", but this evidence is scattered and does not fully substantiate the claim that the use of homœopathy can lead to a reduction in the use of scarce health care resources. Therefore, alongside research into the efficacy and effectiveness of homœopathy, a programme for the economic evaluation of homœopathy needs to be initiated and implemented. As in conventional medicine, a staged approach is considered to be appropriate and described below:

Stage 1. Early economic evaluation / mapping the area

The objective of this stage is to assess homœopathy's potential to offer value for money. Main aspects of this stage are:

- Systematic reviews of the literature
- Establishing in which areas conventional medicine is less than completely effective and mapping this "effectiveness gap" against the evidence with regard to the efficacy and effectiveness of homœopathy. This will enable the identification of the areas where homœopathy is most likely to offer value for money.
- A study of public and private health insurance systems in different European countries with the aim of identifying suitable models for investigating health economic aspects of incorporated (or to be incorporated) insurance policies for homœopathy.

Stage 2. Maturing economic evaluation:

The objective of this stage is the identification of parameters to which the cost-effectiveness of homœopathic treatment is most sensitive by means of pilot studies and sensitivity analyses. The main aspects of this stage are:

- Focused, uncontrolled studies in the most promising areas
- Retrospective and prospective studies in collaboration with private or public health insurance companies.
- Sensitivity analyses with regard to uncertain parameters and the use of analytical modeling techniques.

In the final stages of economic evaluation, clinical and economic data can be obtained from mono- and multi-centre randomised trials to explore the cost-effectiveness of homœopathy when used in routine clinical practice. However, in the coming five years, the emphasis should be placed on stages 1 and 2 as described above.

In this regard, the "Report on NHS practice-based homœopathy project" (ISBN 1901262006) can be used as a reference.

Homœopathic pathogenetic trials

Homœopathic pathogenetic trials or "provings" are a type of research, in which compounds are tested on healthy human volunteers in order to observe as many effects as possible, at a non-toxic level. This careful experimentation is a prerequisite for identifying the true pathogenetic picture of a homœopathic medicine, i.e. its capacity to alter the state of health. The homœopathic pathogenetic trials can in fact be considered as a type of phase-1 study.

Testing of potential remedies on volunteers and describing the induced symptoms in order to establish the "remedy picture" has been performed since the very start of homœopathy. Hahnemann performed the first systematic study of drug action in the history of medicine, and this method, so essential for evaluating the indications for homœopathic remedies, has continued ever since although over time the procedures have changed. Currently, new protocols are under development which are in accordance with modern pharmaceutical and Good Medical Practice (GMP) procedures, for testing new substances or retesting incompletely tested old substances.

It is of the essence that homœopathic medicines are tested on human beings instead of animals. The reason for that is that disease has two distinct forms of expression, to wit the tissue changes or objective signs, as well as the subjective symptoms which include types of pains, emotions and other sensations. Especially these subjective symptoms are of paramount importance because they help to individualise a patient in their illness and thus to select a homœopathic remedy that matches this individualised state of illness.

A recent review of homœopathic pathogenetic trials has shown that nearly 150 such trials, of a wide range of substances, have been conducted in Europe since 1945. The methods used, however, have varied considerably and therefore there is a need for methodological development to refine and standardise the methodology of homœopathic pathogenetic trials. One of the main objectives of the ECH is to establish a "gold" standard for homœopathic pathogenetic trials.

Apart from improving the methodology of homœopathic pathogenetic trials several additional activities should be stimulated such as:

- establishing a priority in retesting currently used remedies.

Some information in old standard homœopathic texts and databases is probably not completely reliable and needs to be verified.

- establishing a priority policy on what kind of new remedies are to be evaluated.

Without doubt, testing the large number of plants, minerals and animal species existing in the world which are not yet used as remedies could increase the therapeutic possibilities of homœopathy even further.

- extracting valuable information from an evaluation of toxicological data (evidence-based
  clinical toxicology), of side-effects of remedies, etc.

In various studies accidental exposures to toxicological compounds such as insecticides, heavy metals, etc. have been carefully recorded. This information can be of relevance either to validate or to extend remedy pictures of compounds used within homœopathy.

- developing a policy on what kind of symptoms should be inventoried or studied in homœopathic pathogenetic trials (biochemical changes in the body, physiological and emotional alterations, types of dreams, etc), as well as establishing their relevance in relation to self-recovery processes.
- establishing the hierarchical order of symptoms.

In this respect it is of importance to differentiate between the quantitative and qualitative character of symptoms. A symptom which is induced by a substance in the majority of healthy volunteers may offer less information than a specific symptom which is induced in a minority of individuals but which may refer to the essence of the tested substance.

- establishing a database and a network for exchange and evaluation of data from observational studies, etc.

For the various mentioned activities a more systematic approach is to be developed.

Homœopathy in the livestock farming sector

In the livestock farming sector homœopathic medicines may replace antibiotics, hormonal and other drugs in some cases of infections or inflammatory diseases or reproductive disorders or may shorten the duration of an antibiotic treatment. Homœopathic dilutions from 10^{-6} M onwards will cause either no residue at all or negligible residues, which can be calculated to be in the low ppb range for worst case scenarios. Therefore, the development and introduction where possible of homœopathy in the livestock farming sector would be particularly important with a view to the more effective protection of consumers from pharmacological residues in animal products and better conditions for livestock currently being reared, in view of the rules soon to be introduced concerning organic livestock farming techniques. Research in this field is necessary.

Auxiliary studies

This research field relates to items or phenomena on which homœopathic treatment is based or which are frequently observed within homœopathic practice, such as "self-recovery", "placebo effect", "semantics", "symptom aggravation", "syndrome shift". All these research fields may in the long run be required to obtain an improved understanding of all aspects of homœopathy as well as its relation to mainstream biomedical knowledge. It should be mentioned that these phenomena are probably not exclusive to homœopathy, but may also be observed in mainstream medicine as well as in other forms of complementary medicine.

FUNDAMENTAL RESEARCH

Constraints to the acceptance of homœopathy which are most often cited by opponents are the lack of a scientific basis and the absence of a theoretical model for homœopathy. In order to improve our knowledge of the fundamental basis of homœopathy (the similia principle and specific preparation of remedies) and to increase the understanding of the working mechanism of its remedies, fundamental research is a prerequisite.

The HMRG recommends the replication of previously used model systems in which an effect of high potencies has been claimed. Although it is recognized that for the validation of high potency studies it is of importance to repeat successful claims in simple model systems by different workers in well controlled (multi-centre) trials, the ECH favours an intensification and extension of fundamental studies. Also in this field of research the 'black box' should be opened.

If one considers the fundamental studies within homœopathy, the majority (>90%) of research is concentrated on demonstrating an effect of potentised agents or remedies. Hardly any research is directed towards the similia principle nor are activities concentrated in a structured programme to unravel the underlying mechanism required to explain the effect of high potencies or to analyze the circumstances for optimizing their effect. From a strategic point of view the existence of an imbalance between research aimed at demonstration of effects and explanatory research is far from optimal, especially when the (non) occurrence of effects cannot be explained in a satisfactory way. Some of these issues will be addressed in this volume.

Fundamental research into homœopathy can be divided into two main fields, namely the *similia principle* and the *preparation* and *mechanism of action* and effect of potentised remedies

The similia principle

Curative approach:

The similia principle implies that remedies or conditions that cause symptoms when applied to healthy biological systems, are used to treat the same symptoms in diseased biological systems. For this curative approach any biological system (cell, organ, plant, animal, human being) can be used in a diseased/ disordered state. The stimulation of recovery by any compound applied according to the similia principle can then be studied. The essential question to be answered is whether the degree of stimulation of self-recovery by low doses is related to the degree of similarity. For research purposes the similia principle can be divided into two main parts, a homologous and a heterologous part. In the homologous part the identical compound is used to disturb the biological system and subsequently, in a lower dose, to cure it. In the heterologous part different substances or drugs (which may either have or not have analogous effects) are used for disturbance and subsequent cure of the biological system.

The similia principle cannot be readily studied in one single type of experiment, since multiple aspects have to be analysed and demonstrated either in a parallel or in a sequential manner. Therefore a research program is required using a biological model system that allows the systematic unravelling of the various aspects involved. In the full analysis of the similia principle a number of steps can thus be discerned. As an example these steps are described at the cellular level, but they can also be applied to more complex systems, e.g. organs and organisms (plants and animals).

Step 1. Selection of parameters for self-defence and -recovery which will be evaluated.

The effect of various compounds or toxins on normal undamaged cells is studied analogous to the description of the symptoms induced by different compounds at higher system levels known within homœopathy as *"provings"*, *"remedy pictures"* or *"homœopathic pathogenetic trial"*.

Step 2. The homologous part of the similia principle (also called 'iso-pathic' approach).

The question is asked whether self-defence and self-recovery, which have been induced by disturbing a cell culture with a sublethal damaging agent, can be further stimulated with a low dose of the identical damaging substance/condition.

Step 3. The heterologous part of the similia principle (specificity of low dose stimulation)

The studies on specificity require the results of steps 1 and 2 and include:

- "provings" of different compounds at the system level under study.
- development of methods to determine the degree of similarity or non-similarity between the symptom pattern of the disordered/ diseased system and the remedy picture of the proven compounds. Based on the degree of similarity the effectiveness of a certain compound to stimulate self-recovery may then be predicted.
- determination of stimulation of self-recovery by low doses of heterologous compounds (ranging from similar (analogous) to non-similar). With this step the above-mentioned prediction can be verified and the question can be answered whether the degree of stimulation correlates with the degree of similarity.

Preventive (c.q. protective or prophylactic) approach.

Although the similia-principle is mainly used as a therapeutic (curative) guideline, it is also used in a prophylactic (preventive) sense. Essentially the same steps as in the above mentioned curative approach are to be taken, i.e. the definition of the parameters of tolerance, the stimulation of tolerance/desensitization to a compound in high dose by a previous incubation with a low dose of the identical compound, and the specificity of the stimulation of tolerance by low doses of analogous compounds. In this field of research the relation with 'hormesis' and 'the adaptive response' should be further established especially in relation to the aspect of specificity.

Potencies vs. dilutions in the study of the similia principle

In essence, the similia principle can be studied without the use of (high) potencies (the founder of homœopathy, Hahnemann introduced the potencies ten years after he first described the similia principle). Thus, for the above mentioned research strategy to verify the similia principle both normal dilutions as well as low or high potency preparations, preferably in a potency range, of different compounds can be used.

Preparation, mechanism of action and effect of potentised substances in various biological systems

Preparation of potentised substances

An important aspect of homœopathy is the specific way in which remedies are prepared (potentised). Potentisation includes a number of different protocols. Nevertheless, there is insufficient consensus yet with respect to the type of potency to be used in different conditions, how to store potencies, which chemical, physical or informational model is most appropriate to explain their mechanism of action, etc. Needless to say increasing the knowledge of these aspects is crucial to improve the quality and the stability of the potencies used.

Mechanism of action of potentised substances

It is suggested that potentisation changes the characteristics of the substance in such a way that an informational content of the remedy is of increasing relevance especially in high potencies. For research purposes a rough division into low and high potencies can be made, related to the way in which an effect might be explained.

- Low potencies, i.e. substances which have been diluted and potentised only a small number of times (for instance D12 or C6 and below). Since these remedies contain a large concentration of molecules of the original substance in the milli- to picomolar range, an effect by low potencies may be explained in terms of biochemistry and current biomedical terminology,
- High potencies, i.e. substances which have been diluted and succussed beyond Avogadro' s number (for instance D24 or C12 and above), are unlikely to contain molecules of the original substance. An effect of high potencies requires additional knowledge, including physics and information theory, in order to explain their action. Some physical and information theory models are presented in this volume. There will be a transitional range (between D12/C6 and D24/C12) where the informational aspect may become more important than the molecular aspect. The claims made for the very high, so-called ultra-molecular dilutions, (i.e. dilutions which, according to Avogadro's Law, are very unlikely to contain even a single molecule of the starting substance) commonly used in homoeopathy are very challenging and raise fundamental scientific questions. Until quite recently it was possible to argue that these extreme dilutions had no real effects, and that all the apparent clinical effects of homoeopathy were due to placebo or non-specific effects. However, the growing evidence of their effects from rigorous, randomised controlled trials is making such a position increasingly untenable.

A possible explanatory hypothesis for the mode of action of extremely high dilutions, or so called ultra high dilutions) is the Information Medicine Hypothesis, which proposes that the actions of homoeopathic medicines should be understood in terms of physically stored information rather than in chemical terms. The Information Medicine Hypothesis states: "Under certain circumstances, water (and perhaps other polar solvents) are capable of receiving and storing information about substances with which they have previously been in contact and of transmitting this information to presented biosystems". If verified, this hypothesis would represent an important scientific advance, with implications extending far beyond homoeopathy. A large part of this volume addresses the problems of the Information Medicine Hypothesis.

Effect of potentised substances

Specific model systems and parameters in the field of chemistry as well as physics are required to study the claims that remedies which have been prepared in a specific way indeed show specific characteristics. This field of research mainly concerns the demonstration of an informational content of high potencies by evaluation of: a) biological effects, and b) structural aspects of potencies.

a. informational aspects of high potencies as measured in biological systems

This research is aimed at determining any effect in any biological organism (ranging from bacteria to man) on the application of high potencies.

b. informational or structural aspects of (high) potencies as measured by physical methods

This research is aimed at determining any change in structure of the solute in potentised remedies. Especially as changes in the structure of water and the importance of an electromagnetic field or frequencies are topics of speculation, physical methods which are able to analyse these aspects should be used or developed to further our knowledge in this field of research.

Priorities in systematic research on high potencies:

• First, research should be focussed on the validation of model systems with which a demonstration of an informational content of the potency is possible using biological and/or physical methods. In this respect, reproducibility in other laboratories is important. A number of model systems may be selected for these studies.

• Second, when a model system has been clearly defined in which an effect can be repeatedly demonstrated, the following questions can be tackled:

- is there a variability in effect due to the different steps in preparation (material and the type of potentiation procedure)?
- is the stability of the potency influenced by electro-magnetic fields, temperature, duration of storage, etc.?
- can the information in the potency be copied?
- what is the cellular/organism device which functions as an informational receiver?
- how can the storage and transmission of information be explained (complex system theory, quantum physics, etc)?

Auxiliary studies

This research field relates to an explanation of items or phenomena which are frequently observed in homœopathic practice, such as "self-recovery", "placebo-effect", "symptom aggravation", "syndrome shift". Furthermore it also relates to fundamental issues such as "hormesis", which is the paradoxical phenomenon that a toxic substance becomes a stimulating agent at a lower concentration. This non-specific stimulation of a large number of physiological processes including growth and longevity by low doses of any toxic compound or stressful conditions has attracted the attention of homœopathic practitioners to serve as an explanatory model for homœopathy. Finally, for an explanation of the high potency effect, physical phenomena such as turbulence and chaotic processes in the preparation of homœopathic remedies, the structure and "memory" of water and/or electromagnetic fields may be of relevance.

All these research fields may in the long run be required to obtain an improved understanding of all aspects of homœopathy as well as its relation to mainstream biomedical knowledge. For an explanation of these items, a system theoretical point of view may be of relevance as well as specific requirements to study complex biological systems.

TARGETS AND PRIORITIES

Homœopathy cannot be demonstrated in one single type of experiment. Multiple aspects have to be analysed and demonstrated either in a parallel or in a sequential manner. Therefore, the ECH stresses the need to develop research programmes that allow the systematic unravelling of the various aspects involved in clinical as well as in fundamental research.

Clinical research

A number of research fields can be discerned, ranging from effectiveness and efficacy research, cost-effectiveness and safety studies to homœopathic pathogenetic trials.

Effectiveness and efficacy research requires a number of sequential steps. Effectiveness research should range from simple data collection in daily practice via observational studies to outcome and cost-effectiveness studies which adhere to strict protocols. Efficacy research focuses on specific questions using the randomised controlled research designs.

The ECH considers the following steps to be essential:

surveys in daily practice

post-marketing surveillance

observational studies

randomised unblinded comparison studies

randomised clinical trials: including both the development of new trials

the replication of well-performed trials.

Research on different aspects of clinical homœopathic practice should be further stimulated, such as the different steps in the diagnostic process, the various steps leading to the selection of a remedy and the development of treatment protocols in different diseases.

From a socio-economic perspective the following studies are necessary:

- demographic studies (who goes to see a homœopathic doctor and why)
- satisfaction and quality of life studies
- economic evaluation of homœopathic health care (cost-effectiveness)

Homœopathic pathogenetic trials (provings) are necessary for a further improvement of the knowledge about existing homœopathic remedies and the quality of homœopathic treatment but also to widen the range of remedies used.

- a "gold" standard for this kind of trial is to be developed
- different compounds need testing and retesting
- toxicology studies are to be carried out and a database is to be developed

The development and introduction where possible of homœopathy in the livestock farming sector would be particularly important with a view to the more effective protection of consumers from pharmacological residues in animal products and better conditions for livestock currently being reared.

Various phenomena described within homœopathy such as initial aggravation, syndrome shift, disease substitution, and placebo effects need further research.

Fundamental research

To improve the knowledge of the scientific basis of homœopathy, a number of activities are to be undertaken:

- A shift in emphasis is required in fundamental research activities towards a more balanced distribution of research activities between the two fundamental aspects of homœopathy, i.e. the similia principle and the use of potentised remedies.
- A formulation of a theoretical model for homœopathy is required with respect to the regulatory processes underlying the stimulation of recovery by application of the similia principle in terms of mainstream biomedical knowledge.
- The relation between the similia principle and hormesis, the paradoxical stimulatory effect of low doses of toxic substances, should be further analysed.
- A limited number of promising model systems in which an effect of high potencies has been claimed, should be reproduced. In this respect it is desirable that two or three conceptually sound test systems should be selected and further developed.
- The conditions in which the optimum effect of (high) potencies occurs in biological model systems should be further analysed.
- The importance of physical phenomena such as turbulence and chaotic processes in the preparation of homœopathic remedies should be further analysed. Furthermore, it should be determined whether the structure and "memory" of water and/or electromagnetic fields are of any relevance in the establishment of an effect of (high) potencies.

Part of this article has been published in May 1997 by the European Committee for Homœopathy as report to the *Directorate General XII of the European Commission*, The editors are grateful to the European Committee for Homœopathy and to all contributing authors.

Footnotes

1. effectiveness = the extent to which a specific intervention procedure, regimen or service does what it is intended to do for a specified population, when deployed in the field in routine circumstances.

2. efficacy = the extent to which a specific intervention, procedure, regimen or service produces a beneficial result under ideal circumstances.

BIO-INFORMATION BETWEEN QUANTUM AND CONTINUUM PHYSICS. THE MESOSCOPIC PICTURE

Jurgen Schulte

Experimental research into structure formation and structure conservation in water and (ultra high) diluted aqueous solutions, as they are used in homoeopathy, have generated a significant amount of interest over the past two decades. Since our last review of academic research into this field (Endler and Schulte, 1994), a noticeable improvement in the scientific standard of experiments, and their publications, has been observed.

Although many experiments are focused on the validation of structure formation and information storage in associated liquids such as water (Endler and Schulte 1994; Demangeat *et al.* 1992), there are only few attempts to develop numerically reproducible model theories based on existing experimental data. This article presents a brief critical review of physical model theories and tries to elucidate their common features and their relation to recent experiments.

Naturally, experiments are based on certain working hypotheses. So far, however, hypothesises seem to be many steps away from representing tenable physical or otherwise scientific theories; they cannot deliver quantitatively reproducible results. Furthermore, they are unable to delimit phenomena, to provide quantitative predictions, or to facilitate generalizations.

Homoeopathy and ultra high dilution research are inherently multi-disciplinary. It is thus no surprise that over the past ten years many very different attempts have been made to establish some understanding of processes that may be interpreted as information transfer or information storage. A promising development of theories has only occurred in the last few years. The scientifically more serious attempts range from earlier, general theoretical observations, based on well-known core concepts of

J. Schulte and P.C. Endler (eds.),
Fundamental Research in Ultra High Dilutionand Homoeopathy, 45-68.

theoretical physics though sometimes of purely philosophical nature, up to recently developed direct approaches, which are more geared toward being experimentally reproducible.

The first section of this article gives a brief overview of possible mechanism of structure formation in liquids in general, and in aqueous solutions in particular, and explains the concept of cluster formation and information storage. In the second section, several established theoretical approaches regarding structure conservation in aqueous solutions are introduced and discussed. A promising theory, namely that of a mesoscopic model of quasi-particles, which has the potential to be quantitatively verifiable, is introduced in the third section. The model tries to bridge the gap between the purely quantum mechanical treatment of interacting molecules and phenomena occurring at the mesoscopic scale. Readers who are less interested in mathematical and physical detail are advised to omit this section. In the last section, theoretical models of dynamical structures in water and experimental findings are discussed in relation to the mesoscopic model of quasi-particles and other models known in literature.

The fundamental riddle of ultra high dilution research is the apparent information transfer and information storage in aqueous solutions as well as in biological cells with aqueous encapsulations. The general perceptions of information storage lead to the easy comprehendible image of information being stored in some sort of *order* or *structure* in the fabric of the information carrier, ie. the aqueous solution or biological cell. Starting with this image, we follow the historical path of physical models which sometimes were originated in completely different disciplines within the physical sciences. As the academic research community became more interested in this fascinating interdisciplinary chapter of research more sophisticated models evolved and it seems some converging of understanding can be observed.

Fundamental principles of structure formation in liquids

Experimental indications of structure formation in aqueous solutions and in biological cells and membranes have long been known, and have been published in recognized academic journals (for comprehensive lists of literature see Endler and Bastide). The idea of information storage in aqueous solutions is briefly outlined in the following standard ansatz in condensed matter physics.

We start with the equation of state for condensed matter, the so called van der Waals equation of state,

$$\left(P+\frac{a}{V}\right)-(V-b)=RT\ .$$

The van der Waals equation establishes a relationship between the temperature T of a system, the occupied volume V, and the current pressure P, in consideration of the volume taken up by molecules alone (represented by the constant a) and the molecular repulsion occurring at small molecular distances (leading to the non-accessible volume b). The constants a and b represent characteristic physical properties of the particular condensed matter (liquid, solid). Most simple microscopic matter can be described to be held together by mutually attractive molecular forces between molecules whose binding energy is proportional to r^{-6}, where r is the distance between two molecules. Atomic nuclei, as well as some quantum-mechanical properties of the atomic shell (electrons), represent a repulsive force which for instance prevents two atoms from collapsing or fusing under normal conditions. The attractive force which causes the binding energy between molecules to fall off with r^{-6} with increasing distance, includes also terms of pure quantum mechanical nature. This force, of course, is of infinite range for all practical purposes, however, its effective range extends to approximately two to three lengths of interatomic distance for most atoms and molecules, and in some cases, another 10-50 molecular distances. Thus the impact of the attractive force of atoms is initially limited to a rather local region of about 10^6Å^3 volume (or 10^{-24}m^3), even if it is in principle of a long-range nature.

For this reason, atoms in liquids tend to "see" only their immediate surroundings, which are made up of approximately 6 to 12 directly neighboring atoms or molecules ("atomic short-sightedness"), and which may include a further several hundred or even a thousand atoms in the vicinity of several additional atomic distances depending on the type of the surrounding molecules. Because of the internal, atom type specific, microscopic structure of a molecule, each molecule has a characteristic electronic-geometric appearance or form when viewed from the outside. This microscopic molecular structure, ie. the atomic electron shell configuration, essentially determines what the geometric configuration of atoms in their very short range vicinity will look like. There is some flexibility within the local structure or spatial arrangement that molecules will assume. The flexibility comes from the constant balancing between potential (binding) energy and the thermal kinetic energy available to the molecules. Hence, the entire liquid system is subject to flexible local geometries. Within certain narrow limits the local geometries can be influenced by external forces, eg. through *succussion* and by change of temperature. The process of succussion is illustrated in Figure 1. The capacity of atoms in the liquid state of aggregation to form local flexible geometric structures, which can be detected by way of experiments and quantum-mechanical calculations, has become the basis for many theories developed over the past years because it is thought that if some of the structures form globally and remain stable over a long period of time, or can replicate themselves serve as a vehicle for potential information storage. The long-term goal of those theories, of course, was to describe the formation of stable macrostructures based on the local and global characteristics, as well as to be able to interpret and predict experimental results. Those geometrically easy to visualize theories, in the following called *geometric theories*, helped to build an important bridge between the community of physical sciences and the physiologist, biologist, and medical scientist who was not accustomed to the complex language of the physical sciences. In the following we summarize the basic concepts of a few prominent geometric theories.

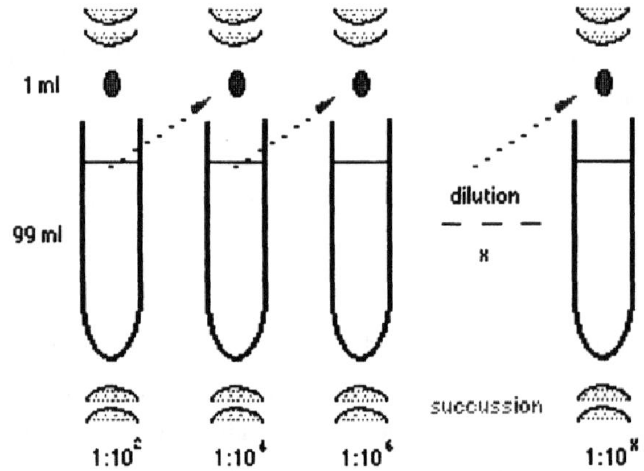

Figure 1. Subsequent dilution and succussion process of an original quantity of a pure substance

In the preceding section the basic mechanism of structure formation in liquids was described rather schematically. A main problem in developing and communicating physical models in this field of research has been the interlink of the macroscopic world and its microscopic substructure, in other words the link between *quantum mechanics* and the measurable continuum *observables*. The various scales of physical forces or dynamics are illustrated in the following example of liquid water. Water and aqueous solutions play a special role among liquids, since the constituting molecule (water, H_2O) is made up of a very large, heavy and polar atom, namely O^- (negative oxygen ion), and two very small, light and polar atoms, namely H^+ (positive hydrogen ion). What difference does the size or weight of an atom make? In the case of heavy atoms, quantum-mechanical effects on the atomic motion (of the core ion) are in principle no longer detectable due to the atoms' physical mass (large *de Broglie* wavelength compared to its atomic size). Thus oxygen may be regarded as a classical object and may be treated according to classical (Newtonian) dynamics. Hydrogen, on the other hand, is the element with the lowest atomic mass, the *de Broglie* wave being of the same order of magnitude as hydrogen's atomic expansion. In water we therefore find objects which follow the principles of classical physics (e.g. the above-mentioned Van der Waals-equation), and objects which should only be described in terms of quantum mechanics. The hydrogen atom of the water molecule looks viewed from the outside like a bare atom core, i.e. a naked proton (for an illustration see Endler and Schulte, 1994). The proton exhibits an almost isolated positive charge which likes to bind with strong negative charges (such as O^-). This special configuration of the hydrogen atom in the water molecule is the reason for the formation of hydrogen bonds in water and associated liquids. The three significant factors which influence local molecular configuration of water are: i) the spatial expansion and classical interaction of oxygen atoms among each other, ii) the inherently quantum mechanical nature of hydrogen atoms, and iii) the strong polarization within water molecules, which facilitates the formation of hydrogen bonds, and which has a significant influence on an extended structure formation. The hydrogen bonds give the water molecule some extra degree of freedom in water, and some extra constraints, to form ordered bonds with other water molecules. The bonds are, however, not as strong as intra molecular bonds, and hence can be broken or arranged with only little energy.

The interaction of short-range quantum-mechanical forces and classical (continuum) forces makes it very difficult to make a general statement about the compactness of such global structures. General experience, however, which is based on quantum-mechanical calculations, lets one suspect that large extended structures might well develop. Whether those extended, or global, structures generate the thermally and mechanically stable self-similar structures necessary for any information storage is yet to be experimentally verified.

The fact that local and some global structure can be formed has been discussed in the vast amount of literature on water and associated liquids. The following section discusses some prominent views on how a local structure as carrier of information may contribute to the formation of global structures. Initial attempts to develop a fundamental understanding of structure formation and related information storage were mainly based on geometric theories.

Formation of distinct global structures

The formation of distinguishable, characteristic global structures in water is a fundamental prerequisite for the possibility of information storage in liquid water. Just as DNA-strands show a well defined stable structure and sequence of macromolecules, and serve as a storage site for complex information, it should also be possible to store information of a certain complexity in aqueous solutions. The hypothesis behind the geometric theories is that information is stored in directly observable variations of local or global structures.

Early theories concerning the possible formation of stable global structures in water were mostly based on somewhat physical-philosophical approaches, as can be found for instance in the earlier work by Popp (1990, 1994). In approaches of this kind general ideas about a collective interaction by way of "*wave functions*" were used to stimulate the imagination with respect to experimentally observed phenomena. However, modern descriptions regarding the formation of characteristic stable structures in water have already left behind those early very handwaving approaches, and it is expected that, with advancing progress of the work in this field, a first theory with concrete, quantitative results worth discussing will give a further motivation to research into structure conservation in liquids in the not too far future.

An explicit approach to global structure formation under the influence of local structure was introduced by Anagnostatos (1994). In this model a local structure, or structure formation, in the liquid causes the successive reproduction of an original structure, hence opening an apparent pathway of information replication. The local structure are thought to have been developed through an externally (electromagnetically) induced fluctuation, or through a deliberate contamination. The structure in the form of a cluster (Anagnostatos 1994) leads to the formation of a local order in the cluster's immediate vicinity and may, via further thermal fluctuations, form so-called clathrates (expanded clusters enclosing smaller cluster clathrates inside), which are reproductions of an original structure. This model of structure formation allows a once implemented structure (information) to disintegrate after a certain (short) life time due to thermal fluctuations, while a present information is preserved by the preceding replication through subsequent clathrates formation. The process of clathrate replication takes place via the formation of a clathrate-typical local order which is thought to be preserved for long enough such that an imprint of its geometry may be left behind in the cluster's immediate vicinity. In this geometric picture the molecular imprint (as clathrate) manifests itself in characteristic bond angles and relative clathrate bondlengths, an effect that ought be observed with standard spectroscopic techniques. So far, characteristics of clathrates have not been experimentally observed as common constituents in aqueous solutions or other associated liquids.

Berezin's approach (1994) to structure formation and structure conservation is closely linked to the extended structures which may develop as a consequence of hydrogen bond formation. In the Berezin model the extended network structure of hydrogen bonds receive a further degree of freedom for information storage (in addition to the information already contained in the hydrogen bond network) via isotopic diversity (isotopicity). In pure water 33 different isotopes can be distinguished, which contribute a large variety of characteristic vibrational frequencies, and thus may stabilize local characteristic features in an extended, seemingly unstable network of hydrogen bonds. This means that global structures can be thought of as the result of joined-together substructures, where redundant information modules may well be present. In this sense, global network structures do not represent the totality of an information unit, but are only the outward manifestation of the underlying isotopic coding. It is not possible to directly influence the "isotopic coding" by simple, electromagnetic means. However, an indirect influence is thought to be possible by electromagnetic coupling onto the hydrogen bonds. Berezin's model leaves the question open how an initially "blank" sample of liquid is encoded with some seed information, say from a deliberately implanted single specimen atom. The "energetic" information that this atom carries relative to the "blank" liquid would have to propagate throughout the liquid causing a cascade of spatial rearrangement of relative isotope locations, an effect which should be observable in a change of the specific heat of the liquid and maybe even in some temperature shift as energy is required for the restructuring process. Such effect has not been yet reported in experiments.

Berezin's model, as well as Anagnostatos' preceding one, can be qualitatively and quantitatively verified by way of the simulation technique of molecular dynamics well known in both physics and chemistry. Thus the usefulness and validity of models and theoretical approaches can be tested numerically in a reproducible way. Such reproducibility should not be confused with a verification of an existing mechanism of information storage because those, and similar models, represent an instantaneous information storage and the question of thermal and temporal stability goes beyond the validity of the respective models.

Preparata and Del Giudice (Del Giudice et al. 1988; Preparata and Del Giudice 1994, and in Taddei 1997) showed that in condensed matter electromagnetic fields can be "trapped" in coherent regions, and that the state of matter containing coherent regions is more stable than that of matter without coherent regions. Coherent regions, which can occupy a spatial region of some micrometers (10^{12}Å^3, or $10^{-18}m^3$), can give rise to condensation by way of the coordinated oscillation of matter components (molecules) and accompanying electromagnetic fields. Through the formation of regions with electromagnetic *coherence* (electromagnetic waves with well defined relative phase), condensed matter can for a long time remain in a state which is more stable, when compared to the formation non-coherent region. Del Giudice et al. showed that the collective motion of water molecules can cause higher polarizability than an equivalent incoherently moving system (Del Giudice et al. 1988). External electrical fields can create metastable extended polarization fields in a coherently moving system of water molecules. In principle it should therefore be possible to create very low frequency long-living polarization fields in water, or to modulate already existing coherent fields, by way of an external field (contaminating substance etc.). The model theory by Del Giudice et al. is similar to the well known formation of magnetic domains found in ferro-magnetic materials, a phenomenon discussed in many standard solid state physics textbooks. As Preparata's and Del Giudice's calculations have shown, the fundamental picture of coherently acting magnetic domains may be extended to domains of coherently acting electromagnetic regions. The major difference between the ordinary textbook treatment and the system we are interested in is that the textbook derivation of domain formation is based on solid state physics assumptions, ie. a rigid frame of domains, whereas in a liquid state such domain boundaries would be flexible according their thermal excitation and may decay within a short period of time. An interesting feature of this model, though, is that it does not necessarily rely on a particular underlying 'seed' geometry, or local architecture, where information may find its vehicle for propagation. The mere existence of a specific coherence phase within domains can serve as information carrier. A possible formation of such domains and their stability can be investigated using Molecular Dynamics simulation on a quantum-mechanical level, a somewhat more challenging undertaking than its ordinary classical application mentioned above (see e.g. Schulte 1996, and the literature cited there in). In their original geometric theory Preparata and Del Giudice's included the element of a dynamic coherence, opening the pathway to further development of well based theories with numerically accessible dynamic features. In this volume Del Giudice and Preparata present recent developments of their earlier work which now allow for a complete dynamic (liquid) treatment of information storage in associated liquids. Their extended model is based on the solid ground of quantum electrodynamics which of course includes the vacuum fluctuation. Here, phase locked molecular excitations give rise to stable coherent regions (domains) which in case of water may extend to about 400Å in radius, ie. $8 \cdot 10^{-22} m^3$). Several coherent regions may couple and thus form the basis for a possible vehicle for information transfer. It is interesting to note that the mechanism for information transfer in the Del Giudice and Preparata model is very similar to the one introduced in the *mesoscopic quasi-particle* model presented in the main section of this paper.

A further step away from the geometric theories, Kratky shows (this volume) that the dynamic of aqueous solutions and the stability of organic systems can be described as dynamic *complex systems* with all the features inherent to complex system theory and chaos theory. Kratky shows that a dynamic system can be perturbed or stabilized by a feedback mechanism, forcing the instability to stay on a dynamical attractor trajectory. The dynamical attractor will then maintain a certain converging path toward the system's point of stability, all in a well defined way. This kind of treatment, or way of looking at the problem, implies manifold possibilities for an experimental examination or manipulation of biological systems, as well as a further possible approach to information storage in associated (aqueous) solutions. However, experimental manipulation of a system through stabilization or destabilization does not provide much direct insight into the actual underlying information or its transfer mechanism. Nevertheless, it would provide an experimental evidence of a stable mechanism of information transfer in liquids.

Xu and Bishop developed a method which could make it possible to reconstruct information from the dynamics a thermal system (Xu and Bishop, 1996), and to determine parameters useful for the controlled manipulation of the system. According to Xu and Bishop, all that is needed for extracting information from a dynamic system are several series of pieces of phase space data from which the dynamics of the total system can be reconstructed, and with which it can be stabilized in a certain state by way of a feedback mechanism (eg. Newton algorithm). This procedure can be employed without any explicit knowledge of the dynamics of a particular system, or its physical boundary condition (Shinkrot et al. 1993). It is of course very helpful when a point of reference (a stable point of the system) can somehow be estimated or guessed. Xu and Bishop found that only very small feedback amplitudes are necessary in order to stabilize a chaotic system by way of using the dynamic feedback mechanism, thus supporting Kratky's conclusion regarding the constructive, external stabilization or manipulation of biological systems. The experimental realization of this model is a very challenging one and is yet to done.

A much different approach has been taken by Lu et al. (1996a, 1996b). Lu et al. reported measurements of dipole based water structures with some resemblance of those theoretically predicted by Del Giudice and Preparata. The main difference between the structures discussed by Del Giudice and Preparata and those found in experiments by Lu et al. is that the Del Giudice and Preparata model is based on the assumption of a pure liquid Lu's findings are clearly based on the presence of ions (impurities) or dielectric contamination. In other words, Lu's findings may have implications in the range of homoeopathic high dilutions whereas the Del Giudice and Preparata model is valid only for ultra high diltutions. Lu measured the ultra-violet (UV) absorption (at 190nm wavelength) of NaCl, HNO_3 and NaOH in liquid water at concentrations between 10^{-3}M and 10^{-13}M. A characteristic decrease in UV absorbance was found with similar behavior for all three solutes at concentration below 10^{-6}M. Figure 2 shows a reconstruction of Lu's data. The pure water reference baseline comprises the UV absorption measurements of water with less than several parts per billion total dissolved solids.

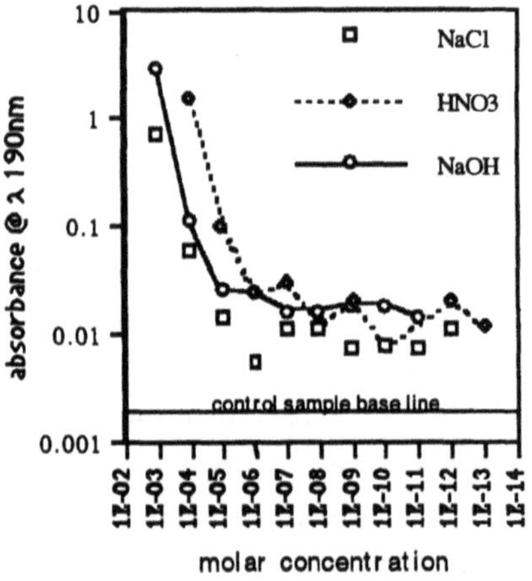

Figure 2: Reconstruction of experimental data from Lu et al.

The interpretation of the experiments is as follows: When an ionic substance is dissolved on water the free ions present strong electrostatic fields. Because of the extremely high dielectric constant of water ($\varepsilon_r \approx 80$), the electrostatic field of ions has a strong impact on the energy density of surrounding water molecules. The first layer of water molecules about an ion will be electrostatically polarized by the ion. The molecules total polarization will cancel, however, if the molecules are arranged spherically symmetric about the ion. Thus, because of Gauss' law, the immediately following next layer of water molecules will experience an energy density boosted by a factor of 80. The energy density , which is energy per volume, is nothing else than the pressure experienced by the water molecules, ie. the water molecules very close to the ion experience an extremely high pressure, causing them to freeze like an ice crystal, only with a very distinct and different crystal structure than ordinary ice crystals. Due to the polarization of the 'frozen' water molecules about the ions, the so formed local rigid structure in water can present thermally very stable dipoles. The pressure about the ion region decreases rapidly (r^{-2}) with increasing distance r from the ion. Note, that this is an electrostatic effect and no electromagnetic radiation (waves) is involved. When two similar charged ions come close together, the pressure on water molecules between them decreases and the spherical ice structure will melt. As the ions move away, pressure builds up again. Thus there will be a continuous process of destruction and growth of highly ordered water structures (crystals). When considering the growth rate of ice crystals and mean collision rate of ions in water (destruction), an ion concentration can be calculated at which the growth rate of crystal structures is larger than the destruction rate. This, of course, can be expected at very low ion concentrations. Assuming a fairly fast growth rate (about cm/s), Lu estimated that the polarized crystals are stable if the ion concentration is smaller than 10^{-5}M, which is considered high dilution in homoeopathic tongue. It is worthwhile to note that Lu chose to prepare the dilutions using subsequent potentization and succusion as illustrated above (Figure 1). Lu et al. also showed that a non-ionic dielectric medium dissolved in water produced similar 'spherical' local ice-structures in, which emphasizes the structure stabilizing effect of polarized region in potentized water. Unfortunately, the dielectric medium in Lu's experiment was not specified, also respective UV absorbance versus molar concentrations were not shown. Also, UV absorbance measurement of potentized unsuccessed water was not shown which could have elucidated the influence of the succussion process.

In Lu's experiments and model, the presence of ions in water, or at least a dielectric seed medium, is essential for growth of well defined highly stable local structures. The model is based of pure electrostatics with the dynamics coming in through ion-ion collisions which destroy the previously stable structures. This, of course, implies that at ultra high dilutions, where we expect no ions to be present any more, the ionic seed mechanism of distinct structure formation seizes to exist, ie. a different mechanism of structure formation or information storage will arise. Furthermore, a transition region will show at very high dilutions where the electrostatic dipole formation gives way to a different mechanism which will no longer rely on strong electrostatics while still requiring a good maintenance of its own distinct seed elements as signifier of physical information. Since the mechanism of information preservation as well as their respective semantic objects changes at the transition between high dilutions ($<10^{-24}$M) and ultra high dilutions ($>10^{-24}$M), it is possible or even likely that the original information content experiences a corresponding transition or change. The experiments by Bastide (this volume) show some indications of a transition of this kind. Unlike in Bastide's interpretation of a possible reaction mechanism in physiological system, ie. the 'non-local sense', non-locality is not required in Lu's model. The signifier can still transport electrostatic and dynamic information the usual way of transport in a physiological system. A long-range non-locality in a physiological system would need to explain why frequent physiological breakdown is not observed, which is an effect that long-range non-locality would cause.

Though in Lu's experiments with the dielectric medium as seed elements or signifiers, measurements still require a more specific discrimination and data presentation, the qualitative results so far hint toward a dielectric medium as possible signifier at concentrations in the ultra high dilution region. The dielectric medium would then be the aqueous solution itself with ordered polarized dielectric seeds (regions, similar to those introduced by Del Giudice and Preparata), ie. a local assembly of molecules that maintain a stable polarized structured despite the thermal dynamics. This can only be achieved if the collision between structures would not lead to their total destruction.

In comparison to earlier approaches, the ones mentioned here have advanced so much by now that standard methods of theoretical physics and theoretical chemistry can be applied, thus providing comprehensible theories which eventually can be verified experimentally and may help in the design of conclusive, well reproducible experiments.

Linking Quantum and Continuum Physics - The Mesoscopic Quasi-particles

The quantum theory, our best theory of physical reality, is actually a theory, not of physical *things*, but of physical *information*. Even today not every physicist would accept this point of view. Or, as John Wheeler articulated it at a conference on quantum mechanics: 'Tomorrow we will have learned to understand and express all of physics in the language of information' (Wheeler 1989). It is perhaps only in the closing stages of the information century that information could be conceived of as having reality, just as in the nineteenth century energy came to have a kind of reality (Milbourne 1996).

The fascinating region between the quantum dimension (10^{-10}m) and the world where the subtleties on quantum level play a minor role (> 0.1μm), is where nano-technology and mesoscopic physics reigns. It is where quantum physics and continuum physics shake hands, in many different and surprising ways. This is the region where we find carbon to be purple, gold to be black, and perfect conductors becoming semiconductors and insulators. There is no magic involved in this, it is a mere play of the laws of physics. Macroscopic properties of condensed matter (such as liquids and solids) are e.g. conductivity, permittivity, compressibility, and other properties that can be measured by macroscopic means. Microscopic, or quantum mechanical, properties are eg. the emission and absorption of photons, rotational and vibrational excitation of molecules and other properties that emerge from the fundamental laws of quantum mechanics. Some of the microscopic properties remain "visible" in bulk condensed matter, though with minor variations, e.g. absorption of photons or electromagnetic waves. Other microscopic properties such as the molecular electron shell configuration adjust according to quantum mechanical laws, yielding a global macroscopic effect such as e.g. electrical conductivity. The modern techniques of spectroscopy reveal accurate fingerprints of quantum mechanics embedded in condensed matter. Spectroscopy uses quantum objects (photons, electromagnetic waves) to probe quantum properties of a material. The wave length of a probing photon prescribes the size of the object it can interfere with or reveal, hence it is able to resolve characteristic properties of molecules and crystal structures in solids. When it comes to larger (mesoscopic) objects such as extended structures in liquids, spectroscopy becomes rather unreliable and researchers resort to indirect measurements and interpretation, often of very limited validity. Nevertheless, it seems possible to bring microscopic fingerprints into relation to show their overall role in the manifestation of macroscopic properties.

One fundamental quantum mechanical property is the *de Broglie* wave length. The Broglie wave length is a fundamental property of all physical objects which have mass or energy. It is important to be aware that the de Broglie wave length is inherently a dynamical quantity which depends on the momentum (velocity, energy) of the object under consideration. For instance, if the vibrational frequency of a molecule changes, the de Broglie wave of the constituent atoms change too. An electron or atom at rest has a de Broglie wave length $\lambda \rightarrow \infty$. Since the (microscopic) molecules in (macroscopic) liquids along with their constituent atoms are constantly in (characteristic) motion which we can observe using spectroscopy (absorption spectra), we may try to link microscopic fingerprints and an underlying characteristic motion within the (macroscopic) liquid.

Using standard spectroscopy the wave numbers of translational vibration of water (0°C) have been measured $k_{tr}^{\mathbf{a}} = 60\ cm^{-1}$, $k_{tr}^{\mathbf{b}} = 190\ cm^{-1}$, and $k_{tr}^{\mathbf{c}} = 190\ cm^{-1}$. The measured absorption frequencies arise from characteristic translational and vibrational motion of the atoms that make up the water molecules water, thus the three characteristic coordinates **a**, **b** and **c**, respectively. It can be shown within a few basic steps that the vibrational motion of molecules can be expressed in terms of dynamical geometric objects. In order to keep things simple we assume that the molecular vibration can be expressed as harmonic oscillation. This assumption is similar to describing the molecule as system of atoms hold together by linear springs, a common assumption in fundamental physics and chemistry. The kinetic energy due to vibrational motion is then $E_{kin} = \frac{1}{2}mv^2$, or expressed in terms of its wave representation in de Broglie form $E_{kin} = \hbar^2 / (2m\lambda^2)$. Here, v^2 is the square of the group velocity of the wave representing the motion of the vibrating atom. The motion of the atom is related to its wave length through the de Broglie relation $p = h\,k = \hbar / \lambda$, with p the momentum of the atom. The more explicit form here is $\lambda = \hbar / (mv)$. The volume λ^3 that a standing wave with this wave length occupies is usually much less than the respective molecular volume V_m. In liquids and solids it is found, however, that this may be different and a volume larger than the molecular one is occupied. The general three dimensional case of standing de Broglie waves can be written $\lambda^{\mathbf{a}} \lambda^{\mathbf{b}} \lambda^{\mathbf{c}} > V_m = V_M / N_A$ with **a**, **b** and **c** the coordinates of the three dimensions and V_M and N_A the molar volume and Avogadro's number, respectively. That means, if the intra molecular motion causes an excitation of larger than classical volume, it may interact with similar excited molecules and synchronization may occur leading to a coherent motion within a local region of more than the spatial extension of one molecule. The regions where the de Broglie waves of molecules can couple and synchronize are dynamical excitations. As with many stable and meta-stable excitations these ones too are then called quasi-particles. In this case, it is a quasi-particle existing between macroscopic and microscopic scale, a mesoscopic quasi-particle as it is discussed in Käiväräinen (1989).

Some remarks need to be made at this point to further explain the above introduction of coherent extended regions forming quasi-particles of some stability. Firstly, we assumed harmonic forces between atoms or molecules. If we allow anharmonicity of oscillations the (most probable) internal kinetic energy of molecules will be less than the corresponding potential energy. Consequently, the (most probable) wave lengths $\lambda^{a..c}$ are somewhat smaller than in the harmonic case. The difference in wave length is, however, negligible for all practical purposes, since anharmonicity is usually very small. In practical terms, any anharmonicity will cause the total number of molecules per quasi-particle to be somewhat smaller than the one calculated here. In the above introduction of the mesoscopic excitation we have not considered that there are in general two very distinct types of intra molecular motions. There are vibrations where atoms are "in phase" (*a*) as indicated in Figure 3, and motions where the atoms move "counter phase" (*o*). In solid state physics these type of motions (within an atomic lattice that makes up a crystal) are called acoustic phase (in phase) and optical phase (counter phase). The two terms stem from the dispersion relation that can be found for the two phases (of motion), where the acoustic phase (*a*) corresponds to a low wave number region where classical mechanics can be applied, and the optic phase (*o*) which corresponds to the high frequency region where coupling to photons (i.e. electromagnetic waves) may occur. Now we have two types of fundamental atomic vibrations from which a quasi-particle can arise.

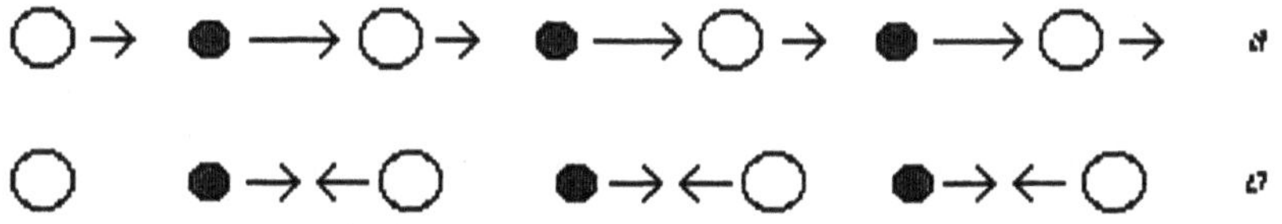

Figure 3. Acoustic (*a*) and optical (*o*) mode of *phonon* vibration.

What happens, if a transition $a \Leftrightarrow o$ between an acoustic and an optic excitation occurs? If there is a region of synchronized excitation, or in other words, an extended quasi-particle exhibiting such coherence, the effect of an $a \Leftrightarrow o$ transition results in the release or take-up (absorption) of energy. As one can imagine such transition will be accompanied by simultaneous density fluctuation causing the propagation of *phonons*. The kinetic energy of an acoustic and optical state are the same. Thus, the energy absorbed or emitted during an $a \Leftrightarrow o$ transition stems from the difference in potential energy of the two states, whereby the a-state has the larger potential energy. It needs to be noted that the vibrational states of molecules are quantized allowing to only vibrate with certain frequencies. Equally quantized then is the energy that can be absorbed or emitted by the quasi-particles. Energy emitted in a $a \Leftrightarrow o$ transition may not only go into production of phonons (density fluctuations) but also into the emission of photons. Similarly, energy required for such transition may come from phonons as well as photons (electromagnetic excitation). The influence of molecules on the properties of the surrounding medium when oscillating in an a- or o-state are much less than the effect of an $a \Leftrightarrow o$ transition, i.e. changing the character of the relative molecular oscillations. The translational and librational bands in the infra-red spectrum of condensed matter ($k_{tr}^{\mathbf{a..c}}$ and $k_{lib}^{\mathbf{a..c}}$) correspond to $a \Leftrightarrow o$ transitions of such local excitations or quasi-particles. In a well coordinated medium such as a protein or membrane we can expect those basic transitions to form further excited states as a result of a coherence of phonons and photons that were generated in such a transition, and we would expect to see further quasi-particles with a much larger intensity than the mere coherence of the original primary excitation would show.

Our initially cited wave numbers $k_{tr}^{\mathbf{a..c}}$ observed in water oatf 0°C are the maxima of the corresponding band spectra. From these maxima we shall derive an expression for the most probable de Broglie waves of quasi-particles. The frequencies ${}^{a}\omega_{tr}^{\mathbf{a..c}}$ of the acoustic state of quasi-particles are

$${}^{a}\omega_{tr}^{\mathbf{a..c}} = \omega_{tr}^{\mathbf{a..c}} / [\exp(\hbar\, \omega_{tr}^{\mathbf{a..c}} / k_B T - 1)],$$

with the frequency and wave number related through the dispersion relation $\omega = \mathrm{v_s}\, \mathrm{k}$ where $\mathrm{v_s}$ is the speed of the wave propagation in the medium (speed of light for photons, speed of sound for phonons). The frequencies ${}^{o}\omega_{tr}^{\mathbf{a..c}}$ of the optical modes are

$${}^{o}\omega_{tr}^{\mathbf{a..c}} = \omega_{tr}^{\mathbf{a..c}} \; [1 + \exp(\hbar\, \omega_{tr}^{\mathbf{a..c}} / k_B T - 1)^{-1}].$$

The term $\exp(\hbar\, \omega_{tr}^{\mathbf{a..c}} / k_B T - 1)^{-1}$ can be interpreted as the average number $\bar{\mathrm{n}}_e$ of excitations having a frequency of $\omega_{tr}^{\mathbf{a..c}}$. Comparing the given relations it can easily be seen than this number increases by 1 for every $a \Rightarrow o$ transition, hence

${}^{o}\omega_{tr}^{\mathbf{a..c}} = {}^{a}\omega_{tr}^{\mathbf{a..c}}\ (\bar{n}_e + 1)$. The librational frequencies can be calculated in a similar way.

The expressions above include the temperature T, and the Boltzmann constant k_B which converts temperature into other common energy units. Through the explicit inclusion of the temperature T we are now in the position to account for characteristic macroscopic temperature dependent properties such as the velocity of sound v_s and the molecular volume V_m. Accounting for the temperature dependence of the velocity of sound (Fine and Millero, 1973) and the molecular volume (Kell 1975) we may plot the de Broglie wave length $\lambda^{\mathbf{a..c}} = (2\pi\, v_{s(T)}) / \omega_{tr}^{\mathbf{a..c}}$ of the quasi-particle as shown in Figure 4. It is notable that the de Broglie wave of the most probable quasi-particle ($\lambda_{tr}^{\mathbf{a..c}}$) exceeds the classical dimension of a single water molecule even at fairly high temperature (Figure 4). How many water molecules do then participate in coherent excitations (quasi-particles)? The concentration of quasi-particles can be calculated from the density of standing waves $\lambda_{tr}^{\mathbf{a..c}}$ (density of states), i.e. $\eta_{tr} = \frac{2}{9}\left(\omega^{\mathbf{a}}\omega^{\mathbf{b}}\omega^{\mathbf{c}} / v_S\right)$, with similar expressions for mean frequencies and librational states. From the density we may derive a quasi-particle volume $V_q = 1/\eta_{tr}$ which then can be used to calculate the number of molecules per quasi particles, $N_m = V_q/V_m$. In Figure 5 we have plotted the number of water molecules per quasi-particle over a range of temperature where we have accounted for the temperature dependence of the molecular volume as mentioned above. Quasi-particles from librational coherent excitations can contain a large number of water molecules (Figure 5) with about 280 at 0°C and even 3 at about 100°C. In the physiological important region of 30°C-40°C we find 70-40 molecules per quasi-particle. This is about the number of water molecules that fit into open protein cavities and occupy a similar amount of volume as protein domains do.

With the introduction of quasi-particles through the fundamental properties of condensed matter , in this case water, a mechanism for long living, metastable, states has been established which opens a pathway for information transfer and storage. Though we have not explicitly mentioned how quasi-particles can actually store information, the reader may have already guessed a possible mechanism.

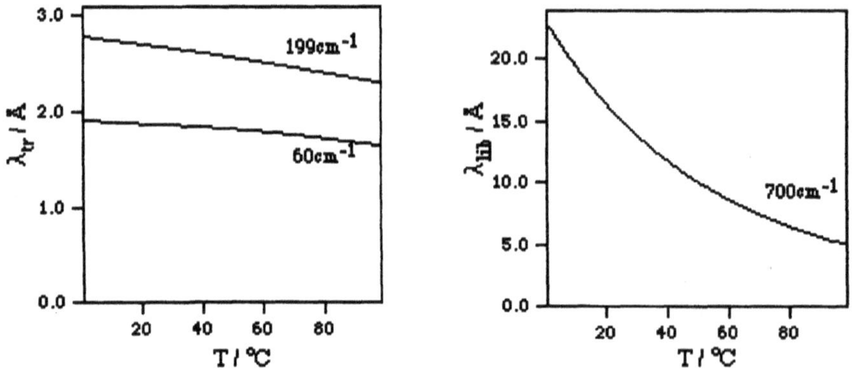

Figure 4. Temperature dependence of λ of most probable translational and librational coherent excitations.

One obvious mechanism is the very existence of an excited coherent state or formation of quasi-particle. The type, energy, and spatial extent of the particle are characteristic to every excitation. In an equilibrium situation (which we find likely in most systems considered in this book) there is an equilibrium (concentration) between $a \Leftrightarrow o$ acoustical and optical transitions in vibrational and librational coherent excitations. If we, for instance, dissolve non-water molecules in liquid water, this equilibrium will be shifted either to the right or left of the previous concentration balance. In the former case the life time of unstable states of a-excitations increase, and in the latter case the stabilization of molecular clusters takes place. If the de Broglie wave length of the dissolved molecule exceeds the spatial dimension of the region of coherent oscillation (the quasi-particle), then it will increase the degree of liquid association. If the de Broglie wave length of the dissolved molecule is smaller than the excited regions, the ordering in liquid structures will decreases. Thus, inert atoms and molecules can have a maximal 'pattern' forming action, stabilizing a-states of excited regions whereas smaller cations will cause the opposite effect. After the contaminating molecules are dissolved in water a new $a' \Leftrightarrow o'$ equilibrium will be established and will be maintained until further perturbation (new molecules added, heat, strong radiation, phase transition, etc.) occurs. That means that even after the "clean removal" of the added non-water molecules the new $a' \Leftrightarrow o'$ equilibrium would be maintained. This new $a' \Leftrightarrow o'$ equilibrium, and respective coherent states, would also be maintained during a clean potentization procedure where the water potentization seed carries the characteristic equilibrium of an original impurity.

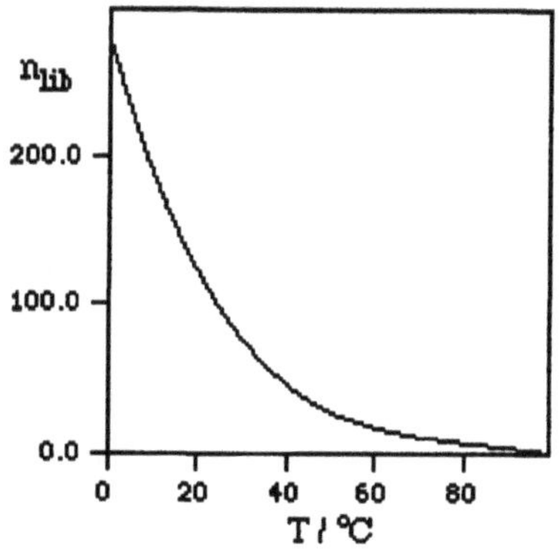

Figure 5. Number of molecules within the volume of most probable librational coherent excitations as function of time.

Shifts in the equilibrium of all types of coherent excitations influence the homogeneity of the dielectric properties and the refraction index on a mesoscopic level. This is caused by the fact that the molecular polarizability in the ground state of the coherent excitations is larger than it is in an excited state (see also Del Giudice and Preparata this volume). The theories of viscosity and conductivity state that such effects induce self-diffusion, non-homogeneity and thermal diffusion. Such processes develop when the spontaneous oscillation between the subsystems of coherent excitations may not necessarily be accompanied by a compensating change in the total internal energy. Thus, long-lasting macroscopic auto-oscillations in condensed matter can be expected in some cases as it has been measured by Chernicov recently (Chernicov 1990). Such system instability is caused by a (discrete) quantum energy distribution competing with continuous thermal energy distribution. Hence, a link can be established between the mesoscopic and macroscopic levels of self-organization.

Mesoscopic Quasi-particles in the context of other models

The formation of hydrogen bonds in water is generally treated as a cooperative process. In the context of statistical thermodynamics, and on the basis of the flickering cluster model, Nemethy and Scherada calculated various thermodynamic parameters, such as free energy, internal energy, entropy as a function temperature, with an accuracy of approximately 3% (Nemethy and Scherada, 1962). Merely the calculation of the heat capacity seemed to be less successful. This effect can be described as a function of temperature due to the decrease in the number of molecules that participate in $a \Leftrightarrow o$ transitions.

A different macroscopic property of water is the magnetic susceptibility χ. The susceptibility χ of water is the result of two contributions, namely that of the mean diamagnetic contribution $\overline{\chi}_d \cong -14.6 \cdot 10^{-6}$ which is induced by an external magnetic field, and that of the mean paramagnetic contribution $\overline{\chi}_p \cong 1.55 \cdot 10^{-6}$, i.e. $\overline{\chi}_{H_2O} \cong -13 \cdot 10^{-6}$. The paramagnetic part is approximately ten times smaller than the diamagnetic part. However, the diamagnetic part is dependent on an external magnetic field and should cease to be detectable as soon as this field is turned off. The number of mesoscopic quasi-particles in water can be estimated to be of order $n \approx 100$ (Figure 5). The interaction energy of quasi-particles with an external magnetic field then becomes $\varepsilon = n\ \mu_B H$. The total energy of quasi-particle interaction with an external field can thus exceed the thermal energy, $\varepsilon > kT$. According to the model, the interaction of an external magnetic field with an *a*-state is stronger (due to phase synchronization) than that with excitations in the *o*-state, $\varepsilon_a > \varepsilon_o$. One consequence of an energetic imbalance is the $a \Leftrightarrow o$ equilibrium being shifted to the left. At the same time the system's potential energy is lowered, since the potential energy U_a, of an *a*-state is lower than that of *o*-state quasi-particle, and the kinetic energies of *a*-state and *o*-state excitations are identical, i.e. $U_a < U_o$, $E_a < E_o$, where E_a and E_o are the total energies ($E_a = U_a + T_a^{kin}$ and $E_o = U_o + T_o^{kin}$. After switching off the external magnetic field, the induced ferromagnetism begins to relax, which may be accompanied by an oscillation of the $a \Leftrightarrow o$ equilibrium. Hence, the quasi-particle formalism seems to be suitable to not only describe spectroscopic features but also electromagnetic properties of condensed matter to which external fields can couple.

In their investigation on biological cell agglomerations Lin, Astumian and Tsong found that, in addition to short-range processes, organic cells also show characteristics of long-range processes, and they implied coordinated organization in the cell agglomeration (Lin et al. 1990). Bellavite and Signorini concluded (this volume) that for thermodynamic reasons such a long-range (and fast) organization of cells requires a signal amplification. They indicated that biological receptors might be a "possible site for this amplification". The assumption was supported by results of experiments directed by Tsong and Lin, which lead to the belief that an enzyme protein can be manipulated in its cell functionality by means of an external electric field.

Long-range electromagnetic interactions among (enzyme) proteins can also be represented in the model of mesoscopic quasi-particles. In this model a "signal amplification" in the traditional sense would not be necessary. The additional entropy, which is necessary for thermodynamic reasons, is provided by the formation of quasi-particles themselves. The water molecules contained in a protein enzyme receptor form a mesoscopic quasi-particle of the size of the spatial expansion of the receptor. When receptor activity occurs, the quasi-particle is activated by translational and electromagnetic vibrational modes, and a direct link with distant quasi-particles takes place. A process that deserves a closer further investigation for it may provide some answers to the (fast) long-range organization of cells and would also serve as benchmark problem for the quasi-particle model.

A sound indication of the necessity of introducing collective coordinates (joint coordinates positioned by a characteristic collective dynamic of an ensemble of particles, or a corresponding subsystem of one of the ensembles) for the description of phenomena in pure water and agglomerations bound by hydrogen bonds (associated liquids) can be found in the work of Cretegny and Peyard (Cretegny and Peyard, 1996). An easily measurable anomaly of water is the unusually high conductivity of ice, just like the mechanism of proton transfer in biological systems (e.g. in DNA-reactions or protein reactions). This phenomenon can also be treated in the context of collective coordination. The charge transport in both systems takes place via defects which build extended structures and include several protons (H^+). Such collective agglomerations with a collective dynamic were already described by Antonchenko, Davidov and Zolotarynk in a physical soliton model (*solitons* are excitations, or propagating waves, which can only develop and exist due to non-linear dynamics of the underlying system; one characteristic property of solitons is that in an ideal case they do not experience refractions or dispersions, and do not interfere with ordinary excitations or waves) (Antonchenko et al. 1983). Based on this approach Cretegny and Peyard developed a further model, in which the original approach is embedded in independent coordinates. In this model chains of hydrogen bonds are described by two substructure lattices, one being formed only by protons (H^+) and one only by heavy ions (OH^- and H_3O^+). The respective centers of sub-lattice defects are then chosen as the collective coordinates. In the context of these collective coordinates the dynamics of the equation of motion can then be solved analytically. The solution to the equation of motion is again a soliton solution (the so called sub-kink-solution), however in his case they are solitons of the collective coordinates, i.e. solitons in each of the two sub-lattices, the solitons being able to interact with each other. The sub-kink-interaction of this mathematical solution implies a possible model for thermal soliton diffusion. In other words, instead of one particle with a (definite) solid mass in an external (periodic) interaction potential (e.g. molecule, cluster, etc.), two quasi-particles with a variable (effective) mass, or mass diffusion, interacting with each other would lead to the same solution. Note that these types of quasi-particles would then be stable for extremely long time.

An interaction of the proton sub-lattice with groups of heavy ions leads then to so called optical modes, characterized by a (protonic) soliton velocity of propagation, which is identical to the velocity of sound of the ion lattice ($v_{soliton} = v_{sound-ion}$). Interactions of the proton sub-lattice with individual heavy ions then lead to acoustic modes with a (protonic) soliton velocity smaller than the ionic velocity of sound, $v_{soliton} < v_{sound-ion}$). Cretegny's and Peyard's model as well as that of Antonchenko, Davidov and Zolotarynk initially refers to extended sub-lattice structures, as they appear for instance in ice. With a few extensions, this model could be stretched to apply also to sub-lattices with internal dynamics, as they appear for example in liquid water. Thus a concept of non-linear domains of excitation (local eigenmodes) could be developed, which would come very close to the above-mentioned model of interacting mesoscopic quasi-particles, a striking similarity worthy of further investigation.

The information transfer of in glass vials sealed homoeopathic thyroxine solution onto the rate of metamorphosis and climbing activity of amphibian larvae, which was introduced to be of non-molecular nature, may also be understood in this context (Endler et al. 1995; Endler et al. this volume). In this case the thyroxine molecule is the original information carrier, however, in the dilution process the information may be transferred onto aqueous solution through the formation of quasi-articles. Photon emissions due to ($a \Leftarrow o$) transitions in the substance inside the vial produce electromagnetic radiation in the infra-red transmission window of the glass, thus stimulating characteristic states in the test basin. In other words, it is possible that the thyroxine molecules form extended, characteristic quasi-particles in their original solution medium, which experience $a \Leftarrow o$ transitions, and which therefore radiate (thyroxine-specific) characteristic photons. The energy of the photons lie in the infra-red or the sub-infra-red region. They can pass the glass material to produce quasi-particle excitations in the external area (basin). It is of course an open question whether the intensity of the radiation is still large enough to produce quasi-particles in the basin.

Conclusion

Physical models have improved considerably over the past decade. Not only have we seen a large variety of well founded attempts to find mechanisms for information transfer and storage in associated liquids, we could also observe an onset of convergence of theories towards only a small number of promising potential candidates for a solid bio-information theory for associated liquids. It is difficult to say at this point which of the proposed mechanism is more likely to succeed. A fair guess, however, is that dynamical theories, including the deterministic chaotic "switching", will play a major role in future research, and some of the dynamics theories may turn out to be not as much different as they seem to be.

In this paper the mesoscopic theory of quasi-particles has been set into context with information storage in high dilutions such as they are found in homoeopathy. The careful reader may have noticed that though the mesoscopic theory is well founded on established physical grounds, with respect to homoeopathy of ultra high dilutions this model still leaves some obvious questions open. Questions that have not been addressed yet are, e.g., how stable and unique the stored information is, how an information transfer or propagation is being kept pure throughout a physiological system, and what is the physiological useful part of the stored information if such excitation can be proved to be stable. Furthermore, if information is transferred from an original chemical substance, can it be synthesised and coded by other means? This touches then a fundamental problem to solve, which is: is the information transfer mechanism the actual information itself or are the information and the information transfer mechanism different as, e.g., in case of a computer disk where a changing magnetic field is the mechanism of information transfer and the sequence of the orientation of the magnetic dipole on the disk represents the information.

Though the physical models work on the very other end of clinical trials or bio-assays, and have still a long way to go, their clear benefit and help in understanding the physics of homoeopathy and ultra high dilutions should be noted. The results from physical models are easily reproducible due to the open books of the laws of fundamental physics as well as may they assist in designing reproducible well founded experiments.

References

Anagnostatos GS, Small Water Clusters (Clathrates) in the Homeopathic Preparation Process; in Endler and Schulte, 1994;121:128.

Antonchenko VY, Davidov AS, Zolotarynk AV, Phys. Stat. Sol. (b) 1983;115:631.

Berezin AA, Ul;tra High Dilution Effect and Isotopic Self-Organization; in Endler and Schulte, 1994;137:169

Chernikov FR, Effect of some physical factors on light scattering fluctuations in water and water biopolymer solutions. Biofizika 1990;35:711.

Chernikov FR, Superslow light scattering oscillations in liquids of different types. Biofizika 1990;35:717.

Cretegny T, Peyard M, Collective coordinate treatment of discreteness effects in hydrogen bonded chains. Phys. Lett. 1996;A21:347-358.

Del Giudice E, Is the "Memory of Water" a Physical Impossibility?; in Endler and Schulte, 1994; 117:120.

Del Giudice E, Preparata G, Vitiello G, Water as a free electric dipole laser, Phys. Rev. Lett. 61, 1988; 1085:1088.

Demangeat JL, Demangeat C, Gries P, Poitevin B, Constantinesco N. Modifications des temps de relaxation RMN à 4MHz des protons du solvant dans les très hautes dilutions salines des silice.lactose. J. Med. Nucl. Biophy. 1992;16,2:135-145.

Endler PC, Pongratz W, Smith CW, Schulte J, Nonmolecular information transfer from thyroxine to frogs with regard to homoeopathic toxicology. Vet. Hum. Tox. 1995;37/3:259.

Endler PC, Schulte J Ultra High Dilution. Physics and Physiology, Kluwer Academic Publishers, Dordrecht (The Netherlands), 1994.

Fine RA, Millero FS, J Chem. Phys 59, 5529ff ,1973

Fröhlich H (eds.), Biological Coherence and Response to External Stimuli, Springer, New York 1988.

Käiväräinen A, J. Mol. Liqu. 41, 1989;55.

Kell GS, J Chem. and Eng. Data 20, 97ff, 1975.

Kozyrev NA, Astronomy and celestial mechanics. In: Problems of investigation of the universe. Moscow, Leningrad 1978;9:582.

Lin DS, Astumian RD, Tsong TY, Activation of Na^+ and K^+ pumping modes of (Na,K)-ATPase by an oscillating electric field. J. Biol. Chem. 1990;265:7260.

Lu SY, Modern Physics Lett. B19, 1996a;909:919.

Lu SY, Lo A, Chong LW, Tianshang L, Hua LH, Geng X, Modern Physics Lett. B19, 1996b;921:930.

Maslovski VM, Postnikov SN Proceedings of the IV seminar on nontraditional technology in mechanical engineering. Sofia-Gorki 1989.

Milburn GJ, Quantum Technology, Allen&Unwin Publishers, Sydney (Australia) 1996.

Nemethy G, Scheraga HA, J. Chem. Phys. 1962;36:13382.

Popp FA, Some elements of homeopathy. Br. Hom. J. 1990;79:161.

Popp FA, Some Biophysical Elements of Homeopathy; in Endler and Schulte, pp, 1994;177:186.

Preparata G, Quanten Field Theory of Superradiance; In: Cherubini R., Del Riaz P., Minetti B. (eds.). Problems of Fundamental Modern Physics. World Scientific, Singapore 1990.

Prigogine I, Strenger I, Order out of Chaos. Hainemann, London 1984.

Schulte J, On the Dynamics of Fullerene Multifragmentation. Supramolecular Science 1996;2:3.

Shinkrot T, Ott E, Grebogi C, Yorke JA, Using small perturbation to control chaos. Nature 1993;363:411.

Taddei-Ferretti C (ed.), High Dilution Effects on Cells and Integrated Systems, World Scientific, Singapore 1997.

Tsong TY, Deciphering the language of cells. Trends Biochem. Sci. 1989;14:89.

Wheeler JA, Information, Physics, Quantum: The Search For Links; Proc. 3rd International Symposium on the Foundations of Quantum Mechanics. 1989:354-368.

Xu D, Bishop SR, Self-locating control of chaotic systems using Newton algorithm. Phys. Lett. 1996;A210:273-278.

BIOENERGETICS AND THE COHERENCE OF ORGANISMS

Mae-Wan Ho

What is the coherence of organisms?

The problem of living organization can be stated as follows:

> *How* is it that an organism consisting of a multiplicity of tissues and cells and astronomical numbers of molecules of many different kinds can develop and function as a whole?
>
> *How* does the organism manage to have energy at will, whenever and wherever required, and in a perfectly coordinated way?

One idea that has emerged over the past 20 years is that it is *coherent*. While the meaning of coherence is unambiguous within quantum theory, difficulties arise when we try to apply the concept to a complex living system with a highly differentiated space-time structure.

The coherence of the organism can most easily be appreciated by a recently developed non-invasive technique that allows one to see the whole organism down to the details of the molecules that make up its tissues. Brilliant interference colours are produced by recombining plane-polarized light split up into slow and fast rays on passing through birefringent liquid crystalline regimes (Ho 1993, 1995).

The principles involved are the same as those used in identifying mineral crystals in geology. Different tissues appear in different colours and varying in intensity according to the orientation and birefringence of the molecules involved as well as their degree of order. The organism - in this case a *Drosophila* larva about to emerge - is obviously alive. Waves of muscle contraction are sweeping over its body, so one can infer that all the molecular motors and enzymes in the tissues are busily turning and deforming as energy is transformed, so how is it possible that they have a crystalline order? It is most likely because the molecular motions are highly correlated or even coherent. As the frequency of visible light is about 10^{14} Hz, and correlated molecular motions generally less than 10^{10} Hz, the tissues will appear indistinguishable from static crystals to the light passing through so long as the movements of their constituent molecules are highly correlated or coherent.

J. Schulte and P.C. Endler (eds.),
Fundamental Research in Ultra High Dilutionand Homoeopathy, 69-88.

With this imaging technique, one can see that the movements of the organism are coordinated all the way from the macroscopic to the molecular. That is what the coherence of the organism entails: it is thick with collective modes of activities over all space-time scales.

This image also brings out the collectiveness or 'wholeness' of the organism. The *Drosophila* larva - like all other organisms we have looked at, from protozoa to vertebrates without exception - is polarized along the anteroposterior axis, as though the entire organism is one single uniaxial crystal. This leaves us in little doubt that the organism is a singular whole, despite the diverse multiplicity of its constituent parts.

I mentioned that the molecules of the tissues maintain their crystalline order when they are actively transforming energy. The evidence suggests that the crystalline order is *dependent* on energy transformation, so that the more energetic the organism, the more intensely colourful it is, implying that the molecular motions are all the more coherent (Ho 1994a). This is consistent with ultra-sensitive high-speed measurements of contracting muscles which show all the molecular motors cycling in synchronous steps (Iwazumi 1987, Granzier et al. 1987). Similarly, X-ray diffraction reveals that a high degree of supramolecular order is maintained during isometric contraction (Bordas et al 1993). The coherence of the organism is therefore closely tied up with its energetic status. To be precise, it is tied up with the way energy is stored and readily mobilized over all its space-time domains.

The problem I would like to address in this paper is how we can understand the coherence of organisms in terms of energy relationships as revealed by thermodynamics and quantum theory, starting more or less from first principles. Some of the arguments are given elsewhere (Ho 1993b, 1994b, 1995b), though none of them as yet complete or fully elucidated.

The organism is not a heat engine

The first thermodynamic characteristic of an organism is that it is not a heat engine. The organism is to all intents and purposes an isothermal system, which means that strictly speaking, no work can be done by heat transfer, for that would require a temperature gradient. What kind of 'engine' is the living organism? Harold Morowitz (Ho 1995b) considers 4 types of engines. The first three, the Carnot engine, the industrial engine and the fuel cell, are all equilibrium devices. As the first two engines operate by heat transfer, they are ruled out. That leaves the third, the chemical fuel cell, and the fourth, ie. the far from equilibrium machine, both promising candidates for the living system.

The living system is remarkable for its efficiency and rapidity of energy transformation. The first clue to its efficiency is offered by analogy to the equilibrium fuel cell, whose efficiency is given by

$$\eta = 1 - T\Delta S/\Delta U \qquad (1)$$

where ΔS and ΔU are the changes in internal entropy and energy and T is the temperature of the surroundings. One way to be efficient is obviously to generate as little entropy as possible.

Does the living system tend towards the minimum of entropy production and maximum efficiency?

The rate of entropy production σ_p in the living system is equal to the rate of increase in entropy σ_i plus the rate of outflow of entropy σ_o,

$$\sigma_p = \sigma_i + \sigma_o$$

At *steady state*, the first term on the right is zero. However, that does not mean entropy production is minimum. As Denbigh (1951) points out, the rate of entropy production may still be very large if the rate of entropy outflow is large. The rate of entropy production σ_p is only a minimum if energy transfer occurs at quasi-equilibrium, or in far from equilibrium conditions, as described later on.

Of equal importance to the efficiency of the living system is the minimalization of free energy dissipation ΔG, so that the quantity,

$$\Delta G = \Delta H - T\,\Delta S \qquad (2)$$

approaches zero, with ΔH being the change of enthalpy. There are two ways to achieve that. The first, which is well-known and ubiquitous in metabolism, is to couple thermodynamic uphill reactions to the downhill ones, so that the negative free energy changes balance the positive ones. The second, not so well-known, is to couple energy transfer directly, by individual enzyme/protein molecules acting as 'molecular energy machines'. In other words, the energy is never thermalized before it is turned into work. Enzymes and proteins, by virtue of their flexibility and size, can absorb energy from the site where it is released, store it, and deliver it directly by appropriate conformational changes to where it is used. This "conservation of free energy", according to Lumry (1991), is achieved via enthalpy-entropy compensation in different parts of the large macromolecule as it undergoes cooperative deformations and movements involving the whole macromolecule. According to Equation (2), free energy change is the difference between the two terms, ΔH and $T\ \Delta S$, which can therefore compensate for each other when enthalpy and entropy changes are of the same sign. At the appropriate *compensation temperature* T_C, which is generally found to be within the physiological range for many reactions (Lumry 1991), the compensation is exact, and $\Delta G = 0$. Thus, one can see that within the living system, positive entropy production can be linked to the generation of work by increasing enthalpy at the same time.

Enthalpy-entropy compensation and free energy conservation also takes place between different enzyme molecules in multi-enzyme complexes which engage in cooperative movements to channel metabolites in sequential reactions without releasing them into the 'bulk aqueous phase' (see below). By "dynamic matching of conformational fluctuations", the collective motions of the associated proteins are no longer independent, but become correlated as a whole (Lumry 1991). This reciprocity in energy relationship will be especially favoured in the quasi-crystalline array of proteins in the membranes of the mitochondria and the chloroplasts, and also in the dense arrays of molecular motors in muscle. However, as I shall show later on, what is being conserved is not 'free energy' but *stored* energy.

Energy storage in the living system - to equilibrate and not to equilibrate

The living system is maintained far from thermodynamic equilibrium; because of that, its temperature does not uniquely define the energy content. Some people argue that the 'real' temperature of the living system must be thousands of degrees Kelvin, but another way to describe the living system is that it has a very high *heat capacity*, or capacity for energy storage. Living systems store a great deal of energy, and both energy storage and efficiency of energy transformation are intimately linked. It is that which enables organisms to adopt the most efficient modes of working in both equilibrium and non-equilibrium regimes (Ho 1993b, 1994b, 1995b).

An organism is nothing if not organized heterogeneity, with nested dynamic structures over all space-time scales. The differentiation of the body into organs, tissues, and cells is familiar to everyone. The cell itself is partitioned into many compartments by cellular membrane stacks and organelles that can respond directly to external stimuli and relay signals to other compartments. Within each compartment, *micro*-compartments can be separately energized to give local circuits; and single enzyme proteins, or complexes of two or more proteins can function as molecular energy machines that cycle autonomously without immediate reference to its surroundings (Ho 1995b).

Spatial differentiation in the living system, therefore, spans at least ten orders of magnitude from Å (10^{-10} m) for intramolecular interactions to metres for nerve conduction and the general coordination of movements in larger animals. The relaxation times of processes range from $<10^{-14}$ s for resonant energy transfer between molecules to 10^{7} s for circannual rhythms. Something is surely missing from any account that treats the living system as though it has a single, homogeneous 'steady state'.

The physiologist Colin McClare, who was involved in reformulating thermodynamics so that it could be applied to individual molecules, and not just to statistical ensembles of molecules, first introduced the important notion of a characteristic time of energy storage (McClare 1971). This characteristic interval of time t, at temperature q, partitions the energy of the system into *thermal energies* that reach equilibrium in a time less than t, and the *stored energies* that remain in a non-equilibrium distribution for a time greater than t.. So, stored energy is any form which does not thermalize, or degrade into heat in the interval t.

The explicit introduction of time, and hence *time-structure* enables us to see that there are two quite distinct ways of doing useful work at maximum efficiency in the living system, not only slowly according to conventional thermodynamic theory, but also quickly - both of which are reversible and generate little or no net entropy. (This is implicit in the classical formulation, $\Delta S \geq 0$, for which the limiting case is $\Delta S = 0$. But the attention to time-structure makes much more precise what the limiting conditions are.)

A slow process is one that occurs at or near equilibrium. The efficiency as measured by Equation 1 approaches its maximum 1 as ΔS approaches zero. By taking account of characteristic time, a reversible thermodynamic process merely needs to be slow enough for all thermally-exchanging energies to equilibrate, ie, slower than t, which can in reality be a very short period of time for processes that have short time constants. The effect of spatial partitioning - from compartments to micro-compartments - is to restrict the volume within which equilibration occurs, thus reducing the equilibration time. This means that *local equilibrium may be achieved for many biochemical reactions in the living system*. We begin to see that thermodynamic equilibrium itself is a subtle concept, depending on the level of resolution of time and space.

At the other extreme, there can also be a process occurring so quickly that it, too, is reversible. In other words, provided that the exchanging energies are not thermal energies in the first place, but remain stored, then the process is limited only by the speed of light. Resonant energy transfer between molecules is an example of a fast process. It occurs typically in 10^{-14}s, whereas the molecular vibrations themselves die down, or thermalize, in 10^{-9} s to 10^{1} s. Resonant energy transfer is 100% efficient and highly specific, being determined by the frequency of the vibration itself. This process is now known to be involved in the primary steps of photosynthesis where energy transfer and electron transfer occur with great speed and with almost 100% quantum yield (Fleming and Grondell, 1985). Resonant energy may also be involved in muscle contraction as McClare has suggested. Recent evidence indicates that energy from a single molecule of ATP may be delocalized over 4 or more work cycles of the molecular motor(s) (Yanagida et al. 1985).

McClare restated the second law so that it could apply to single molecules, say, enzyme molecules acting as molecular energy machines: *useful work is only done by a molecular system when one form of stored energy is converted into another. In other words, thermalized energy is unavailable for work and it is impossible to convert thermalized energy into stored energy.*

McClare's restatement of the second law of thermodynamics is unnecessarily restrictive, and possibly not valid, for thermalized energy from burning coal or petrol is routinely used to run machines such as generators and motor cars. Furthermore, when one takes the nested compartmental structure of the living system into account, then thermalized energies from a small compartment will still be contained within a larger encompassing compartment, so there is a possibility it may be available for work.

For example, enzymes embedded in a membrane which can undergo cooperative correlated motions could channel thermal energies to enzymically active conformational changes. In other words, local temperature fluctuations within an isothermal system may perform work, which is not contrary to the second law, for the living system is not at thermodynamic equilibrium.

I suggest a more adequate restatement of the second law as follows (Ho 1993b, 1994b, 1995b): Useful work is done by molecules by a direct transfer of stored energy, and thermalized energy cannot be converted into stored energy in the same system. A 'system' is here defined by the extent to which thermal and other exchanging energies equilibrate within the relaxation time of the process involved. This also clearly demands a more specific definition of the spatial extent of equilibration, which will be given below.

Energy storage over all space-time domains

It is of interest to compare the thermodynamic concept of 'free energy' with the concept of 'stored energy'. The former is strictly an ensemble concept, it cannot be defined *a priori*, much less can it be assigned to single molecules, as even changes in free energy for an ensemble cannot be defined unless we know how far the reaction is from equilibrium. Thus, Lumry's "free energy conservation" is strictly speaking, *stored* energy conservation. Stored energy, as defined by McClare with respect to a characteristic time interval, can be extended to a characteristic spatial domain, so one can generalize the concept of *energy stored within a characteristic space-time*. Stored energy therefore, depends explicitly on the space-time structure of the processes in the system, and is relevant it to whole organisms as well as to single molecular machines (Ho 1995b). For example, energy is stored as bond vibrations or mechanical/electrical strains in protein molecules within a spatial extent of 10^{-9} m to 10^{-8} m and a characteristic time scale of 10^{-9} s to 10^{-8} s. For a human being, the overall energy storage domain is in metre-decades. In between these two extremes, energy is stored in nested spatio-temporal compartments which are locally in equilibrium, but globally out of equilibrium with respect to one another, with equilibration space-times spanning the whole gamut between the local and fast to the global and slow.

How is energy mobilized in living systems?

Energy is mobilized in living systems by coupled flows of metabolites. The flow of metabolites is coupled to a flow of electrons and protons up and down the electronic-protonic gradients via the interconversion of ATP and ADP. The energy of the photon absorbed by chlorophyll is coupled to electron transport. Electron transport is coupled to the translocation of protons across the energy transducing membrane. The proton gradient thereby created supplies the protonmotive force for the synthesis of ATP from ADP and Pi. And finally, the hydrolysis of ATP back to ADP and Pi is coupled to practically all thermodynamically uphill or energy requiring reactions. All coupled flows are vectorial, the flows are in the direction of their respective forces or gradients. Two features of the coupled flows are notable.

First, the couplings are *symmetrical* for the most energetically efficient processes. It means that the forces have reciprocal effects on the coupled flows, and also, if the forces are reversed, so are the flows. This applies to ATP synthesis from ADP and Pi, coupled to proton transport in oxidative and photosynthetic phosphorylation; as well as ATP splitting coupled to the molecular motor in muscle contraction. ATP is split into ADP and Pi by the ATP synthase embedded in the membrane when the proton gradient is run in reverse, just as ATP is synthesized by the molecular motor when ADP and Pi are supplied.

The second notable feature of the coupled flows of energy and material is that they are *cyclical*, as a casual glance at a metabolic chart will convince us.

Cycles differ in lengths from the tricarboxylic acid cycle of core metabolism to the relatively short redox cycles in the elements of the electron transport chain and the two state interconversions of intermediates such as NADH/NAD and ATP/ADP. Are the two features - symmetrical coupling and cyclical flows - predicted from thermodynamics?

The thermodynamics of symmetrically coupled flows

The quasi-equilibrium approximations of the steady state developed by Onsager (Onsager 1931) show how symmetrical coupling of linear processes can arise naturally in a system under energy flow. A system of many coupled processes can be described by a set of linear equations,

$$J_i = \Sigma_k L_{ik} X_k \quad (3)$$

where J_i is the flow of the ith process (i = 1, 2, 3.....n), X_k is the kth thermodynamic force (k = 1, 2, 3,.....n), and L_{ik} are the proportionality coefficients (where $i \neq k$) and coupling X_k. Onsager showed that for such a multi-component system, the couplings for which the X_ks are invariant at microscopic level with time reversal (i.e., velocity reversal) will be symmetrical; in other words,

$$L_{ik} = L_{ki} \quad (4)$$

The main difficulty in applying Onsager's result to the living system is that the latter is far from thermodynamic equilibrium and operating in the nonlinear regime, whereas Onsager's reciprocity relationship is only valid for the linear regime close to thermodynamic equilibrium. However, Onsager's reciprocity relationship has recently been generalized by Sewell (Sewell 1991) to *nonlinear processes exhibiting space-time scale invariance*. Those are the characteristics of a whole class of critical phase-transitions (Bruce and Wallace, 1989) that may well include the living system (see later). Symmetrical coupling will apply for as long as those coupled processes are dispersion free, and hence stable.

An archetype of such critical phenomena is the Bénard convection cells that arise in a pan of water heated uniformly from below. At a critical temperature difference between the water at the top and the water at the bottom, bulk flow begins as the less dense, warm water rises from the bottom and the denser, cool water sinks. The whole pan eventually settles down to a regular honeycomb array of flow cells. So long as the temperature difference remains, the cells are stably maintained as heat flow couples (symmetrically) to the bulk flow of water.

The condition of dispersion-free macroscopic observables is satisfied in a pure phase, which, as Sewell points out, is a prerequisite to any deterministic law including that of Onsager. Sewell's generalization of the Onsager reciprocity relationship applies to locally linearized *combinations* of forces, which nonetheless behave globally in a nonlinear fashion. This is particularly relevant to the living system, where nested compartments and micro-compartments ensure that many processes may be operating locally at thermodynamic equilibrium even though the system as a whole is far away from equilibrium. Also, as each process is ultimately connected to every other in the metabolic net through catenations of space and time, even if truly symmetrical couplings are localized to a limited number of metabolic/energy transducing junctions, the effects will be shared or delocalized throughout the system, so that the reciprocity relationship will apply to appropriate *combinations* of forces, precisely as formulated by Sewell.

Another important assumption which justifies the application of Onsager's relationship to the living system is that suggested by Denbigh (1951). It is to regard the system in question as a superposition of dissipative and non-dissipative processes, so that the Onsager relationship applies only to the latter. In other words, it applies to coupled processes for which the net entropy production is zero,

$$\Sigma_k \Delta S_k = 0 \qquad (5)$$

This will include most of the thermodynamic processes in living systems because of the ubiquity of coupled *cyclic* processes, for which the net entropy production is zero, as expressed in Eq. (5).

The thermodynamics of cyclical flows

The other important development in the thermodynamics of the steady state came from Morowitz, who derived a theorem showing that at steady state, the flow of energy through the system from a source to a sink will lead to at least one cycle in the system (Morowitz 1968). The proof goes as follows.

For a canonical ensemble of systems at equilibrium with i possible states, where f_i is the fraction of systems in state i (also referred to as occupation numbers of the state i), and t_{ij} is the transition probability that a system in state i will change to state j in unit time. The principle of microscopic reversibility requires that every forward transition is balanced in detail by its reverse transition, ie,

$$f_i \, t_{ij} \neq f_j \, t_{ji} \qquad (6)$$

If the equilibrium system is now irradiated by a constant flux of electromagnetic radiation such that there is net absorption of photons by the system, i.e., the system is capable of *storing* energy, a steady state will be reached at which there is a flow of heat out into the reservoir (sink) equal to the flux of electromagnetic energy into the system. At this point, there will be a different set of occupation numbers and transition probabilities, f_i' and t_{ij}'; for there are now both radiation induced transitions as well as the random thermally induced transitions characteristic of the previous equilibrium state. This means that for *some* pairs of states *i* and *j*,

$$f_i' t_{ij}' \neq f_j' t_{ji}' \qquad (7)$$

For, if the equality holds in all pairs of states, it must imply that for every transition involving the absorption of photons, a reverse transition will take place involving the radiation of the photon such that there is no net absorption of electromagnetic radiation by the system.

This contradicts the original assumption that there is absorption of radiant energy (see previous paragraph), so we must conclude that the equality of forward and reverse transitions do not hold for some pairs of states. However, at steady state, the occupation numbers (or the concentrations of chemical species) are time independent (ie, they remain constant), which means that the *sum* of all forward transitions is equal to the *sum* of all backward transitions, ie,

$$df_i'/dt = 0 = \Sigma(f_i' t_{ij}' - f_j' t_{ji}') \qquad (8)$$

But it has already been established that some $f_i' t_{ij}' - f_i' t_{ji}'$ do not balance. That means other pairs must also be non-zero to compensate. In other words, members of the ensemble must leave some states by one path and return by other paths, which constitutes a cycle. Hence, *in steady state systems, the flow of energy through the system from a source to a sink will lead to at least one cycle in the system.*

Coupled energy flows are symmetrical and cyclical

The two results - Onsager's reciprocity relationship and Morowitz' theorem of chemical cycles - I believe, imply a third: that *symmetrically coupled cycles* will arise in open systems which are capable of storing mobilizable energy under energy flow (Ho 1994b). What are the thermodynamic consequences of symmetrical coupling and cyclic energy relationships?

Let us take cycles first. Cycles return to the same point, and hence the net entropy change is always zero (c.f. Equation (5) above); and little or no entropy *accumulates* in the system. These are the relevant non-dissipative processes for which Onsager's reciprocity relationship will apply. More importantly, *cycles can be subject to coherent coupling*, and that may be why living processes are universally organized over a range of 'biological rhythms'. *Coherent* coupling, which I shall say more about later, is not why we can move the whole body together, but why we can move different parts independently!

Why is *symmetrical* coupling important? Because it allows energy to delocalize over the whole system *as well as* to localize to any point, which is ultimately why we can have energy at will, whenever and wherever required. However, to achieve the rapidity with which energy is mobilized in living systems, one requirement is that the energy stores must be distributed over all space-time scales, as it indeed appears to be. For example, skeletal muscles are rich in ATP, whose concentration remains constant, as it is rapidly replaced by creatine phosphate. Before the latter is used up, muscle glycogen breaks down to supply ATP from glycolysis. In the longer term, the lactate accumulating has to be cleared away by the blood supply in exchange for glucose from breaking down glycogen stores in the liver. The various energy stores are themselves replenished progressively starting from very localized substrate oxidation in the mitochondria. Intuitively, one can see that for the most efficient mobilization, the energy stores have to be distributed evenly over all space-time domains, so that every scale can be readily bridged. This is analogous to the problem of percolation in many length and time scales (Bruce and Wallace, 1989), where large gaps will compromise the transparency of the system.

Long-range energy continua

The organism is indeed vibrant with energy flows on every scale bridging the local and the global, the fast and the slow. Metabolic fluxes are now very actively investigated and evidence is accumulating that the fluxes are dynamically organized in detail down to the molecular level: metabolites are 'channelled' or passed sequentially from one enzyme to the next without being released into the 'bulk aqueous phase'. In effect, the cell is partitioned into numerous metabolic 'micro-compartments' that separate parallel, simultaneous fluxes. Meanwhile, a number of different lines of investigation are converging to the conclusion that perhaps no proteins in the cell are dispersed at random in solution, but are instead, organized in an almost solid state. The 'solid' phase also contains a high proportion of the metabolites, and much of the cell water may actually be bound or structured by its enormous amount of surface area (Welch 1985). This detailed dynamic organization is optimized thermodynamically, in terms of the efficiency of energy transformation, and kinetically, in terms of the speed with which reactions take place.

These conditions are also very favourable for one of the most neglected energy flows in living systems: electricity and associated electrostatic, electrochemical, dipole and electromagnetic interactions that span all space and time scales from the superfast exchange reactions between contiguous molecules and resonant energy transfers to long-range, global electric and ionic currents and electromagnetic signals between cells and organisms. With characteristic insight and foresight, Szent-Györgi has written 25 years ago (Szent-Györgi 1960),

"... life is driven by nothing else but electrons, by the energy given off by these electrons while cascading down from the high level to which they have been boosted by photons. An electron going around is a little current. What drives life is thus a little electric current."

Welch and Berry (Welch and Berry, 1985) argue for "long-range energy continua" connecting all parts of the cell in electrochemical fluxes. In particular, they draw attention to the proton currents (proticity) that may also be flowing, constituting a "protoneural network" that could play a large role in regulating cellular metabolism. Many enzymes, for example, can conduct protons along the hydrogen bonds and/or act as sensors of local electric fields.

The overriding feature of energy mobilization in living systems is that it is *stored* energy that is being mobilized over all space-time scales, for it is stored energy that is capable of doing work. Stored energy is none other than *coherent* energy, and the domain of storage is the coherence domain. This immediately suggests that the living system has a full range of coherence times and coherence volumes, the extent of which far exceeds any other physico-chemical system.

Coherent excitations

The energy efficiency of living systems can be adequately accounted for by the thermodynamic considerations I have outlined so far. However, the rapidity and precision with which the energy is mobilized, to my mind, requires additional explanations, and this brings us to coherence defined both classically in terms of phase-transitions, and more rigorously in quantum theory.

An example of phase transition is the Bénard convection cells already mentioned - a phase transition phenomenon in which random molecular movements are transformed into globally coherent flows.

Another example is the laser, where energy is pumped into a cavity containing atoms capable of emitting light. As the pumping rate is increased, a threshold is reached - the laser-threshold - at which all the atoms oscillate together in phase, and send out a giant light track that is a million times as long as that emitted by individual atoms. The mathematical theory describing collective phenomena such as the laser (and the Bénard convection cells) is of sufficient generality that it predicts the emergence of global order under very different circumstances. Could something similar be involved in the living organism?

The first detailed suggestion for that was presented by Herbert Fröhlich from the late 1960s to the late 1980s just before he died. Similar ideas had been put forward earlier by Schrödinger, Szent-Györgi and Prigogine (Schrödinger 1944, Szent-Györgi 1960, Prigogine 1967).

Fröhlich (Fröhlich 1980) argued that as organisms are made up of strongly dipolar molecules packed rather densely together, electric and elastic forces will constantly interact. Metabolic pumping will excite macromolecules such as proteins and nucleic acids as well as cellular membranes (which typically have an enormous electric field of some 10^7V/m across them). The excited molecules/membranes will vibrate at various characteristic frequencies resulting from the coupling of electrical displacements to mechanical deformations. This eventually builds up into collective modes (coherent excitations) of both electromechanical oscillations (*phonons*, or sound waves in solid medium) and electromagnetic radiations (*photons*) that extend over macroscopic distances within the organism and perhaps also outside the organism. The emission of electromagnetic radiation from coherent lattice vibrations in a solid-state semiconductor has recently been experimentally observed for the first time (Dekorsky et al. 1995). The possibility arises that organisms may actually use electromagnetic radiations to communicate between cells or between different organisms (Ho 1994c).

If that is the case, then, as Fröhlich's theory predicts, organisms will be extremely sensitive to weak electromagnetic fields, perhaps through specific coherent excitations, or by interfering with coherent excitations at phase transition. In my laboratory, we have found that brief exposures of early fruitfly embryos to weak static magnetic fields cause characteristic global perturbations to the segmental body pattern of the larvae emerging 24 hours later (Ho 1992). As the energies involved are well below thermal threshold, our conclusion was that there can be no effect unless the external field is acting on a coherent domain where charges are moving in phase, or magnetically sensitive dipoles undergoing phase alignment globally (Ho 1994d).

Although Fröhlich's theory is far from generally accepted, the concept of coherence is already subsumed, or taken for granted, in the description of many macroscopic biological functions, from the synchronous flashing of light among huge populations of fireflies to the coordination of the movements of the four limbs in animal locomotion. In the latter case, each limb has to be treated as a single oscillator with a well-defined collective phase relationship to the other limbs (Collins 1992). This is an accurate description of what actually happens: each limb moves as one, and not as an unwieldily collection of independent tissues and cells. Similarly, our heart beats as a whole and maintains a phase relationship with our respiratory cycle. Furthermore, as the organs are functioning, the specific groups of nerve cells in the central nervous system connected to the organs will also be firing regularly in unison and exactly in phase with the rhythmic movements of the organs (Kelso 1991). Let us unravel what is involved here: it is assumed - correctly as it turns out - that something as complicated as a limb, or a heart, or a whole respiratory system, nevertheless possesses a collective phase of all its multiplicity of activities, i.e., it is coherent. For only when the subsystem is coherent can it couple coherently to other subsystems. This principle extends throughout the organism's space-time domains over which energy is mobilized - each domain being capable of working as an independent coherent unit that is yet in step with the whole. This is where something like *quantum* coherence may be able to provide some more insight, as I shall explain later.

Fröhlich's theory has been extended by a number of theoretical physicists who show that coherent excitations can arise under the most general conditions of energy pumping and energy sharing, and that *once established, they are stably maintained* (Duffield 1988). This significant result also invites one to identify an maximum principle for the thermodynamics of open systems which is analogous to that of equilibrium systems.

The thermodynamics of organized complexity

In working through the bioenergetic relationships of living processes described so far, I came to the conclusion that the following postulates may form the beginnings of a thermodynamics of organized complexity (a slightly different version was presented earlier (Ho 1994b)):

1. Open systems capable of storing energy will evolve to maximize energy storage over all space-time domains, such that the entropy function (analogous to Gibbs entropy S_G),

$$S_G = -\Sigma_k\, p_k(r,t)\, ln\, p_k(r,t) \tag{9}$$

increases under sustained energy flow, with p_k the space-time population.

2. At a certain threshold of energy supply, a phase-transition occurs at which energy mobilization and storage over all space-time domains are coupled together to a single degree of freedom.

3. At phase transition, energy is effectively stored with equal population over all space-time domains, i.e.,

$$p_k(r,t) = const \qquad (\Sigma\, p_k(r,t) = 1) \tag{10}$$

4. This implies that the entropy given in Equation (9) is both a maximum for the system, but also a minimum at phase transition because the modes are coupled together to a single effective degree of freedom.

The $p_k(r,t)= const$ regime is one of maximum entropy because the *potential* degrees of freedom are maximized over all space-time domains, but it is also the regime of minimum entropy because the activities in all space-time domains are coupled together so there is only a single *actual* degree of freedom (Ho 1994b). Phase transition-like phenomena may be more general than we think. And whenever they occur, something like a maximum-minimum entropy $pk(r,t) = const$ regime may be involved. Recent work on ant colonies has shown that while individual ants exhibit random behavioural patterns, the collective can undergo phase transition to regular periodic behaviour when the number of ants in the colony reaches a certain threshold. At phase transition, there appears to be a maximum of entropy, measured in terms of the number of active ants per unit period of time, and also a maximum of correlation between ants that are active or inactive (Sole and Miramontes 1995). It would be of interest to examine the Fourier spectra at phase transition to see if they too, go through a maximum at phase transition.

The $p_k(r,t)= const$ regime can also be described in terms of the *coherent quantum state* or 'pure' state consisting of a superposition of many coherent states, so that all possibilities are immediately accessible. The adaptability of the organism depends on just this seemingly paradoxical property. For, only by maximizing the potential degrees of freedom is it possible to access the single degree of freedom that is required for coherent action (Ho 1994b).

It should be noted that the $p_k(r,t)= const$ regime is a generalization of the discovery made by Popp from many years of experimentation on light emission from living organisms, which I shall briefly describe below, as it offers further insights into the coherence of organisms.

The coherence of biophoton emission

Although there have been many claims that organisms emit and receive electromagnetic signals in biocommunication, these signals are difficult to detect below the visible range. Fritz Popp is one of the pioneers in detecting ultraweak photon emission from living systems. He and many others since, have found that all organisms emit light ('biophotons') at ultraweak intensities which are strongly correlated with the cell cycle and other functional states (Popp 1992). The emitted light typically covers a wide band (200nm to 900nm) around the optical range - the limitation being usually set by the photon-detecting device - with approximately equal numbers of photons throughout the range (Popp 1986).

Biophotons can also be studied as stimulated emission after a brief exposure to light of different spectral compositions. It has been found, without exception, that the stimulated emission decays according to a hyperbolic function; which Popp and Li argued to be a sufficient condition for a coherent light-field (Popp and Li 1992). This implies that photons are held in a coherent form in the organism, and when stimulated, they are emitted coherently, like a very weak, multimode laser. Such a multimode laser has not yet been made artificially, but it is at least not contrary to the theory of coherence in quantum optics as developed especially by Glauber (Glauber 1969), *so long as the modes are coupled together.*

There is, indeed, evidence that the modes within the visible range are coupled together. Spectral analyses of the stimulated emission show that it always covers the same broad range, regardless of the composition of the light used to induce it, and furthermore, it can retain its spectral distribution even when the system is perturbed to such an extent that the emission intensity changes over several orders of magnitude. Furthermore, the hyperbolic decay kinetics is uniform throughout the spectrum (Muscumeci and Godlevski 1992).

Another evidence for the coherence of the photon (energy) field within each organism is that populations of synchronously developing *Drosophila* embryos can undergo phase-correlated collective light emission minutes to hours after a single brief light stimulation (Ho 1992b). In order to build up such a phase-correlation, each individual embryo must itself be highly coherent with a definite phase that can phase-lock, or couple coherently, to all the others in the population (Zhou et al 1994). That is how the most rapid and effective biocommunication may be achieved in living systems; and we must, finally, consider the important implications of quantum coherence itself.

What is quantum coherence?

In order to begin to understand what *quantum coherence* entails, let us look at Young's two-slit experiment in which a source of monochromatic light is placed behind a screen with two narrow slits. As is well-known, light behaves as either particles or waves according as to whether one or both slits are open. When both slits are open, even single photons behave as waves in that they seem to pass through both slits at once, and, falling upon the photographic plate, produces a pattern which indicates that each photon, in effect, interferes with itself! The intensity I or brightness of the pattern at each point depends on the sum of four correlation functions:

$$I = G(t,t) + G(b,b) + G(t,b) + G(b,t) \qquad (11)$$

where $G(t,t)$ is the intensity with only the top slit opened, $G(b,b)$ the intensity with only the bottom slit opened, and $G(t,b)+G(b,t) = 2G(t,b)$ is the additional intensity (which take on both positive and negative values) when both slits are opened. At different points on the photographic plate, the intensity is

$$I = G(t,t) + G(b,b) + 2|G(t,b)|\cos\theta \qquad (12)$$

where θ is the angle of the phase difference between the two light waves.

The fringe contrast in the interference pattern depends on the magnitude of $G(t,b)$. If this correlation function vanishes, it means that the light beams coming out of t and b are uncorrelated; and if there is no correlation, we say that the light at t and b are incoherent. On the other hand, increase in coherence results in an increase in fringe contrast, i.e., the brightness of the bands. Since $cos\theta$ is never greater than one (i.e., when the two beams are perfectly in phase), then the fringe contrast is maximized by making $G(t,b)$ as large as possible and that signifies maximum coherence. But there is an upper bound to how large $G(t,b)$ can be. It is given by the Schwarz inequality:

$$G(t,t,)G(b,b) \geq |G(t,b)|^2$$

The maximum of $G(t,b)$ is obviously obtained when the two sides are equal:

$$G(t,t)G(b,b) = |G(t,b)|^2 \qquad (13)$$

Now, it is this equation that gives us a description of quantum coherence. A field is coherent at two space-time points, say, t and b, if the above equation is true. Furthermore, we have a coherent field if this equality holds for all space-time points, e.g. X_1 and X_2. This coherence is called first-order coherence because its refers to correlation between two space-time points, and we write it more generally as,

$$G_{(1)}(X_1, X_1)G_{(1)}(X_2, X_2) = |G_{(1)}(X_1, X_2|^2 \qquad (14)$$

The above equation tells us that the correlation between two space-time points in a coherent field *factorizes*, or decomposes neatly into the self-correlations at the two points separately, and that this *factorizability is a sufficient condition for coherence*. Factorizability does not mean that the pure state can be factorized into a mixture of states, but it does lead to something quite unusual - any two points in a coherent field will behave statistically independently of each other. If we put two photon detectors in this field, they will register photons independently of each other.

Coherence can be generalized to arbitrarily higher orders of m, say, to m approaching ∞, in which case, we shall be talking about a fully coherent field. If mth order coherence holds, then all of the correlation functions which represent joint counting rates for n-fold coincidence experiments (where $m<n$) factorize as the product of the self-correlations at the individual space-time points. In other words, if we put n different counters in the field, they will each record photons in a way which is statistically independent of all the others with no special tendency towards coincidences, or correlations (Glauber 1969).

The key to understanding the coherence of organisms is in the *factorizability* of the quantum coherent state. The coherence of organisms entails a quantum superposition of coherent activities over all space-time domains, each of which correlates with one another and with the whole, and yet independent of the whole. It is this factorizability that underlies the sensitivity of living systems to weak signals, and their ability to communicate and respond with great rapidity. It is why we can attend to all the different vital functions simultaneously and separately, and yet remain an undivided whole.

Conclusion

I have approached the problem of living organization by considering bioenergetic relationships in thermodynamics, where I show how some of the main features of energy mobilization in the living system - its efficiency and rapidity - can be explained by symmetrically coupled, cyclical flows of stored energy over all space-time domains. That is where the possibility for coherence emerges as a critical phase transition, thus connecting with Fröhlich's ideas of coherent excitations and finally, with quantum coherence. The thermodynamical description both leads to, and converges with, the description based on quantum coherence. The living system is maximally efficient, communicative, responsive, and most of all, factorizable, in the sense that the maximum correlation of the local to the global is realized simultaneously with the maximum local freedom. When one ceases to see that as a paradox, one has finally grasped the meaning of organic wholeness or the coherence of organisms.

Acknowledgment

It is a pleasure to thank Geoffrey Sewell for explaining his generalization of Onsager's reciprocity relationship to me and for many other inspiring comments. I am also grateful to J. Pokorny for helpful suggestions.

References

Collins J.J., Stewart I.N.J. Math. Biol. 1992;30:827-838.

Bordas J., Diakun G.P., Diaz F.G., Harris J.E., Lewis R.A., Lowy J., Mant G.R., Martin-Gernandez M.L., Towns-Andrews E. Two-dimensional time-resolved X-ray diffraction studies of live isometrically contracting frog sartorius muscle. J. Musc. Res. Cell. Mot. 1993;14:311-324.

Bruce A., Wallace D. Critical point phenomena: universal physics at large length scales. In: Davies P. (ed.). The New Physics. Cambridge University Press, Cambridge 1989:236-267.

Dekorsy T., Auer H., Waschke C., Bakker H.J., Roskos H.G., Kurz H. Emission of submillimeter electromagnetic waves by coherent phonons. Phys. Rev. Lett. 1995;74:738-741.

Denbigh K. G. The Thermodynamics of the Steady State. Malthuen & Co. Ltd., New York 1951.

Duffield N.G. Global stability of condensation in the continuum limit for Frohlich's pumped phonon system. J. Phys. A. Math. Gen. 1988;21:625-641.

Fleming G.R., van Grondell R. The primary steps of photosynthesis. Physics Today 1994;2:48-55.

Fröhlich H. The biological effects of microwaves and related questions. Adv. Electronics and Electron. Phys. 1980;53:85-152.

Glauber R.J. Coherence and quantum detection. In: Glauber R.J. (ed.). Quantum Optics. Academic Press, New York 1969.

Granzier H.L.M., Myers J.A., Pollack G.H. Stepwise shortening of muscle fibre segments. J. Musc. Res. Cell. Mot. 1987;8:242-251.

Ho M.S., Stone T.A., Jerman I., Bolton J., Bolton H., Goodwin B.C., Saunders P.T., Robertson F., Brief exposure to weak static magnetic fields during early embryogenesis cause cuticular pattern abnormalities in Drosophila larvae. Physics in medicine and Biology 1992a;37:1171-1179.

Ho M.W., Xu X., Ross S., Saunders, P.T. (1992). Light emission and rescattering in synchronously developing populations of early Drosophila embryos - evidence for coherence of the embryonic field and long range cooperativity. In: Popp F.A., Li K.H., Gu Q. (eds.). Recent Advances in Biophoton Research. World Scientific, Singapore 1992b:287-306.

Ho M.W., Lawrence M. Interference colour vital imaging - a novel noninvasive technique. Microscopy and Analysis 1993a;9:26. See also Ho M.W., Neural Network World 1995a; 5: 733-750.

Ho M.W. The Rainbow and the Worm: The Physics of Organisms. World Scientific, Singapore 1993b.

Ho M.W., Saunders P.T. Liquid crystalline mesophases in living organisms. In: Ho M.W., Popp F.A., Warnke U. (eds.). Bioelectromagnetism and Biocommunication. World Scientific, Singapore 1994a.

Ho M.W., Popp F.A., Warnke U. Bioelectrodynamics and Biocommunication. World Scientific, Singapore 1994c.

Ho M.W. What is (Schrödinger's) negentropy? 3rd BTK Conference Proceedings, Innsbruck 1994b.

Ho M.W. (ed.). Bioenergetics, Living Processes, An Open University Third Level Science Course, Open University Press, Milton Keynes 1995b.

Ho M.W., French A., Haffegee J., Saunders P.T. Can weak magnetic fields (or potentials) affect pattern formation? In: Ho M.W., Popp F.A., Warnke U. (eds.). Bioelectrodynamics and Biocommunication. World Scientific, Singapore 1994d.

Iwazumi T. High-speed ultrasensitive instrumentation for myofibril mechanics measurements. Am. J. Physiol. 1987;252:C253-C262.

Kelso J.A.S. Behavioral and neural pattern generation: The concept of neurobehavioral dynamical systems. In: Koepchen H.P., Huopaniemi T. (eds.). Cardiorespiratory and Motor Coordination. Springer-Verlag, Berlin 1991:224-234.

Lumry R. Mechanical, force, hydration, and conformational fluctuations in enzyme catalysis. In: Kuby S.A. (ed.). A Study of Enzymes, vol.2, Mechanism of Enzyme Action. CRC Press, New Haven 1991.

McClare C.W.F. Chemical machines, Maxwell's demon and living organisms. J. theor. Biol. 1971;30:1-34.

Morowitz H.J. Energy Flow in Biology. Academic Press, New York 1968.

Musumeci F., Godlevski M., Popp F.A., Ho M.W. Time behaviour of delayed luminescence in Acetabularia acetabulum In: Popp F.A., Li K.H., Gu Q. (eds.). Recent Advances in Biophoton Research. World Scientific, Singapore 1992:327-344.

Onsager L. Reciprocal relations in irreversible processes I and II. Phys. Rev. 1931;37:405-426 and 1931;38:2265-2279.

Popp F.A., Li K.H., Gu Q. (eds.). Recent Advances in Biophoton Research. World Scientific, Singapore 1992.

Popp F.A. On the coherence of ultraweak photoemission from living tissues. In: Kilmister C.W. (ed.). Disequilibrium and Self-Organization. Reidel, Dordrecht 1986:207.

Popp F.A., Li K.H. Hyperbolic relaxation as a sufficient condition of a fully coherent ergodic field. In: Popp F.A., Li K.H., Gu Q. (eds.). Recent Advances in Biophoton Research. World Scientific, Singapore 1992:47-58.

Prigogine I. Introduction to Thermodynamics of Irreversible Processes. Wiley & Sons, New York 1967.

Schrödinger E. What is Life? Cambridge University Press, Cambridge 1944.

Sewell G. Non-equilibrium statistical mechanics: dynamics of macroscopic observables. In: Gans H., Blumen A., Amann A. (eds.). Large-scale Molecular Systems: Quantum and Stochastic Aspects. Plenum Press, New York 1991.

Sole R.V., Miramontes O. Information at the edge of chaos in fluid neural networks. Physica D 1995;80:171-180.

Szent-Györgi A. An Introduction to Submolecular Biology. Academic Press, New York 1960.

Welch G.R. (ed.). Organized Multienzyme Systems. Academic Press, New York 1985.

Welch G.R., Berry M.N. Long-range energy continua and the coordination of multienzyme sequences in vivo. In: Welch G.R. (ed.). Organized Multienzyme Systems. Academic Press, New York 1985.

Yanagida T., Arata T., Oosawa F. Sliding distance of actin filament induced by a myosin crossbridge during one ATP hydrolysis cycle. Nature 1985;316:366-369.

Zhou Y.-M., Ho M.W., F.A. Popp, Q. Gu (1994). An oscillator cellular automaton model of superdelayed luminescence in synchronously developing populations of early Drosophila embryos. In: Beloussov L., Popp F.A: (eds.). Non-Equilibrium and Coherent Systems in Biophysics. Biology and Biotechnology, Moscow (in press).

COHERENT ELECTRODYNAMICS IN WATER

E. Del Giudice and G. Preparata

The phenomenological properties of water are shown to be inconsistent with the generally accepted approach to condensed matter, whose basic assumption is that interactions between molecules are electrostatic and short range. On the contrary, Quantum Electrodynamics (QED) is seen to play a key role in the existence of the liquid state. In particular, a collective dynamics emerges, which could have significant biological implications.

INTRODUCTION

From the very beginning of science, water has been the object of many puzzling questions and intriguing proposals among scientists and natural philosophers. Quite correctly, water was identified very early as the fundamental element for life. The name of Thales remains associated to this truth and the whole further understanding has but confirmed it. Today we know that 99% of all molecules in living matter are water molecules. Nevertheless, the generally accepted scientific descriptions of water still struggle with many unsolved puzzles and mysteries. The basic microscopic element of water is quite well known. Indeed, the molecule of water is accurately described as an ensemble of two hydrogen nuclei, one oxygen nucleus and ten electrons; the minimum energy configuration of this system is well known; and we possess also a remarkably accurate knowledge of the energetically excited configurations: all this wisdom can be easily found in the textbooks (Franks 1972, Neilsons and Enderly 1986). However, our insight of the fabric of liquid water, of the dynamics which transforms an ensemble of small molecules, whose interaction in the gas phase does not exhibit any dramatic feature (notice that at 100C and 1 atmosphere less than 1% of the microscopic components of water vapour are reported by spectroscopic evidence to be dimers or trimers or more, of the basic molecule), into a complex liquid with properties not easily traceable back to the structure of the isolated molecule.

J. Schulte and P.C. Endler (eds.),
Fundamental Research in Ultra High Dilutionand Homoeopathy, 89-103.

The generally accepted theory of water, through a skilful "bricolage" of molecular properties, accounts for only a fraction of the observed phenomena. There is still a "magic of water" - the magnetic response of water, the memory of water, the response of water to the low frequency interval of electromagnetic radiation - which is not explained, not even qualitatively, by this "accepted wisdom". This inability can probably attributed to the stubborn denial by the academic community of the very existence of such "magic" phenomena.

However, there are also "orthodox" phenomena which defy the generally accepted wisdom and we will start just from the orthodoxy in order to challenge this wisdom. In Section 2 we will describe and discuss some of these puzzles, in Section 3 we will expose the points of disagreement between the accepted theory of water and the universally accepted Quantum Electrodynamics (QED), in Section 4 we will give a description of liquid water based on QED. Section 5 will be devoted to the biological perspective opened by this novel approach, while Section 6 will contain some indications useful for a future understanding of the "magic phenomena".

SOME PUZZLING PHENOMENA OCCURRING IN WATER

Almost all existing models of the structure of liquid water are based on the assumption that the interactions among molecules are fully static (a dynamic model of structured coherent regions in water is discussed by Schulte in this volume, an entropic approach is presented by Ho, this volume). Different potentials (Stillinger 1980) have been introduced, some of them not related in any obvious way to the well known electron distribution of the molecule in its ground state and to any reasonable polarization induced by neighbouring molecules (Buckingham in ref. Neilson and Enderly 1986). There is, however, a general consensus on the fact that mesoscopic structures in the liquid, such as H-bonded structures (Stanley 1980), clathrates (Dreyfuss and Wachmann 1981) or clusters of some kind (Schulte 1994), do emerge from the electrostatic forces assumed to exist among water molecules. Moreover, these forces have to be conceived as short-range ones, given the global electrical neutrality of matter and the consequent shielding occurring within matter. The total potential energy of a sample of liquid water should, thus, depend only on the mutual positions of molecules; its time average should be a function of the intermolecular distances only and, consequently, of density. Since in the thermodynamical states along the critical isotherm, density is the same on both sides, liquid and vapor, of the curve, we should expect that, at comparable temperatures, in liquid and vapor quite similar molecular dynamical properties should appear. Where, then, the well known thermodynamical differences between liquid and gas should come from? Are we witnessing a departure from the historical program started last century by Maxwell, Boltzmann, Gibbs and others that sees thermodynamics as the consequence of the dynamics of the elementary components?

In order to shed light into this puzzle it is useful to discuss a recent experiment (Pastorino et al. 1993) performed at Rutherford Lab in the UK. Two specimens of water were analyzed, the first in the liquid sector of the phase diagram, the second in the gas sector. The thermodynamical variables were fixed as follows:

	Liquid	Gas
Temperature	300C	400C
Pressure	95 bar	800 bar
Density	0.72 g/cm^3	0.66 g/cm^3

Notice that in water $T_{critical}$=374C. The difference between the two specimens may be considered small with respect to the strength of an H-bond, which in Kelvin units (k=Boltzmann's constant=1), is about 2800K. So one should expect similar correlation functions g(OO), g(OH), g(HH) in the two cases. Note that recent molecular dynamics simulations reported in Pastorino et al. (1993) just predict similar functions. However, neutron scattering measurements, probing the correlation functions, gave quite different responses in the two specimens. Although after the appearance of the paper of Pastorino et al. there has been an extensive discussion and reassessment of its results, a discrepancy between the two sets of correlation functions still holds, also in the most conservative estimates (Soper et al.). As a result, it appears that the molecules of each specimen happen to know which phase they belong to and to be able to give a response in kind. In other words, at comparable temperatures (a difference of 100 K is small with respect to 2800 K) and densities, the same molecules are seen to produce two completely different structures. In this way, the dynamic state of the system appears to depend on thermodynamics. But who would inform the molecules when to produce a liquid or a gas? By which arcane virtue the same molecules may be able to exhibit different forces in different thermodynamic conditions? The only reasonable view appears to believe that molecules have different structures in different thermodynamic phases. But then what in the world would convince the molecules to restructure themselves at the phase transition? A description of the liquid in terms of localizable molecules interacting through short range forces appears finally to be in big trouble.

Moreover, two different intermolecular organizations are found to coexist in liquid water. A first evidence for this two-fluid structure of liquid water is provided by

Raman and infrared (IR) spectroscopy. The vibrational peaks that in molecular spectra (gas phase) range in the interval 1500-3700 wave numbers, in the liquid spectra appear at frequencies lower than in the gas and depend on temperature T. This feature has been understood (Walrafen 1968) in the following way: consider each large peak in the liquid vibrational spectra as the juxtaposition of two peaks, one corresponding to the unbound molecular spectrum, the other emerging from an ensemble of bound molecules. Since the relative abundances of the ensembles depend on T, the same occurs for the relative intensities of the component peaks, giving rise to the T dependence of the peak of the envelope.

A further indication of the two fluid structure of water comes from the reported existence of a fast sound in water (Teixeira et al. 1985). Sette et al. have carefully measured by inelastic X-ray scattering this "fast sound" dispersion curve $\omega=\omega(q)$ in the range $0.4Å^{-1} < q < 1.4Å^{-1}$ (Sette et al. 1995), finding in this range a slope more than double than the ordinary sound speed both in H_20 and D_20, ruling out a mode propagation involving essentially the hydrogens. The authors interpret this curve as a bend up of the ordinary acoustic branch at ordinary momenta. However, as we have pointed out in Del Giudice and Preparata (1995a), their dispersion curve is remarkably linear in the range $0.35Å^{-1} < q < 1Å^{-1}$, intersecting the origin with surprising precision. Moreover, in the same q-range the experimental values of the width , the $\Gamma(q)$ are perfectly consistent (Fig. 4 in Sette et al. 1995), showing no trace of the quadratic q-dependence typical of the relaxation of the longitudinal viscosity. Also in this case, as in supercritical water, molecular dynamics (MD) simulations fail to reproduce the experimental data. The only reasonable interpretation (Del Giudice and Preparata 1995a) of the "fast sound" is thus that of considering it as the proper sound in the ensemble of bound molecules responsible for the lowest peak in the Walrafen doublet, whereas ordinary sound is the proper sound in the ensemble of unbounded (or less bounded) molecules. The problem thus arises of understanding how a water molecule could be brought to belong to one or the other phase. This was just the objection raised by Bernal and Fowler (1933) to earlier proposals of two-fluid water, such as the one proposed by Roentgen (1892) on the grounds of the observed T-dependence of the solubility in aqueous solutions.

Let us now turn to the "magic", the unorthodox observations on water, ignored by orthodox theorists just because out of reach of any orthodox combination of epicycles around the basic electrostatic interaction among molecules. Apart from the memory of water, whose features are widely analysed in this volume, we would like to discuss briefly the magnetic properties of water. This topics is absent in the academic literature and some material can be found sketchily in technical publications of the involved firms (Eiler Sundt). Devices exploiting the above properties are commercially available and patents have been issued.

The phenomenon is basically the following. Water is kept streaming in a pipe at a speed v and passes through the poles of a permanent magnet, where a field B, orthogonal to v, is maintained. If $|B|$, $|v|$ and the cross-section of the pipe are given appropriate values, remarkable effects are seen to appear. Scales no longer form and, furthermore, old scales are dissolved; corrosion is sharply reduced, surface tension is diminished. It seems that, for normal pipes whose diameter is a couple of inches and a field intensity of some thousands of gauss, the ideal speed is about 2 m/sec. The above phenomenon should be compared to the one reported by Piccardi in 1959, who observed, too, a decrease of surface tension when submitting water to an e.m. field of 10 kHz. It seems that the magic properties of water have been recognized as early as 1890 in Germany (Eiler Sundt). We wonder why no attention, not even negative, has been given by the orthodox "water community" to this phenomenon. Actually a point of view, which depends heavily on models of water in terms of well localized molecules, should be embarrassed by the finding of non trivial magnetic behaviour in an ensemble of molecules whose magnetic dipole moment is just zero.

DYNAMICAL INSTABILITY OF A SYSTEM OF MOLECULES INTERACTING THROUGH STATIC FORCES ONLY.

The phenomena discussed so far suggest that the generally accepted model of liquid water appears to fail in describing those observed effects. Let us briefly summarize such models.

Water molecules are assumed to be mutually connected by H-bonds (Stanley and Teixeira 1980), arising from the hydrogen nuclei of the water molecules, supposedly abandoned by their electrons migrated into the electron cloud of the molecules, become sources of attraction of the electron clouds of neighbouring molecules, which bulge out and form a "bridge" with the hydrogen nucleus. Unfortunately, as shown in Neilson and Enderly (1986), the structure of the isolated unexcited molecule exhibits no resemblance to the above picture. The formation of an H-bond would then require a deep restructuring of the involved molecules and the necessary energy (the strength of an H-bond should be at least 5 kT at room temperature) is just not there. But, at least temporarily, let that be. Suppose we have an ensemble of molecules kept together by short range static forces of some kind. How stable is this system against the vacuum quantum fluctuations?

It is well known that in quantum theory (QED is the only theory able to accurately describe matter so far), the e.m. field has an intrinsic fluctuation, like all other fields, so that a rigorously zero e.m. field is ruled out. Evidence for these vacuum e.m. field fluctuations is provided by the "Lamb shift". In 1948 Willis Lamb was able to account for the small discrepancy between the measured and the calculated (in terms of atomic physics) values of the energy of the hydrogen electron, predicted by the interaction energy between the current of the orbiting electron and the (small) fluctuating e.m. field in vacuum. The Lamb shift is very small, indeed. But what about the shift when the ensemble of atoms or molecules is very dense?

This problem has been tackled in the last ten years and the results of this analysis have been recently published in a book (Preparata 1995). The problem is to ascertain whether reshuffling of the electron cloud, obtained through the excitation of a well defined molecular level, could be made energetically possible by compensating the energy investment necessary for the excitation by means of a "super Lamb shift" term, namely the (negative) interaction energy between the electron current, produced in the reshuffling, and the e.m. field of the fluctuation, which could get thus possibly amplified in the process. Should this compensation occur, a purely static system could become unstable against the fluctuation. As we will show in Section 4, the physical situation corresponding to the coupled fluctuation of the molecules and the e.m. field becomes then energetically advantageous over the original, non fluctuating configuration, since the "super Lamb shift" term provides the "energy profit". The system needs less energy by oscillating back and forth between the two levels selected by the coupling with the e.m. fluctuation field than by staying motionless in the configuration where the field is zero and the molecules are in their individual ground state!

In the new configuration induced by quantum fluctuations, the individual molecules acquire on one hand a collective behaviour and on the other are completely restructured, since their individual configuration is given now by the superposition of two different molecular states; notice that the excited state included in this superposition could be several eV above the ground state whereas one would need tens of thousand to achieve the same result in a thermal way. In particular, the deep restructuring of the electron cloud of water molecules required for the formation of H-bond becomes now feasible through the above non thermal, non entropic quantum mechanism.

THE STRUCTURE OF LIQUID WATER AS DETERMINED BY QED

Let us illustrate in this Section the steps of the phenomenon of the condensation of an ensemble of molecules forming a dilute gas into a much denser liquid. A detailed analysis of the general dynamics of condensation can be found in Ref. 18; the particular case of liquid water is discussed in Del Giudice et al. (1995b) and Arani et al. (1995).

Our aim is to try and find out by what mechanism the general laws of QED succeed in forming the liquid, starting from a bunch of independent molecules that make up the gaseous state. We assume that a large number N of such independent molecules roam in a volume V of empty space. As stressed in the preceding section, this space cannot be really "empty", since it is filled with the fluctuations of all conceivable quantum fields (gravitational, chromodynamic and so on), including the e.m. field. It is just such fluctuations that render eventually unstable, under suitable conditions, the ensemble of N molecules quietly milling around in space.

As a quantum object, each molecule has a discrete spectrum of energy levels. For the sake of simplicity, let us assume at first that the molecular levels are only two, later we will restore the whole complexity of the real molecule. In the following we will use the "natural unit" system, where $h/2\pi = c = k_B = 1$, generally adopted by quantum field theorists. Let us define

$$\omega = E(2)-E(1) \tag{1}$$

the difference of the energies of the two levels. It is possible to show (Preparata 1995, Arani et al. 1995). that the fluctuations of the e.m. modes with frequency w can couple resonantly to the molecules in a region whose minimum size is just the mode wavelength $\lambda=2\pi/\omega$ - inducing a matter field fluctuation. Molecules oscillate between the two levels in tune with the e.m. fluctuation field and, by doing so, produce an electromagnetic current which couples with the electromagnetic field. The total energy of the fluctuations is thus the sum of three terms:

1. the energy of the electromagnetic field fluctuating around the classical vacuum (i.e. the zero-field configuration)
2. the energy carried by the excited molecules
3. the energy of interaction between the molecule current and the fluctuating electromagnetic field.

The relative weight of these terms depends on the density N/V. For very dilute systems, as in the "Lamb shift" case, the third term is negligible, so that the system must borrow energy to keep the fluctuation alive, which, according to the Heisenberg principle, is doomed to fade away. However, the third term has been shown show (Preparata 1995, Arani et al. 1995) to increase with the particle density faster than the first two.

There is then a critical value of the density beyond which the total energy of this fluctuation becomes negative. The fluctuating e.m. field starts thus a runaway evolution since the negative energy makes it unstable and as a result its amplitude increases. At the same time, the molecules' density becomes larger, for it is in this way that the combined system proceeds towards the state of minimum energy, the ground state. In such state the ensemble of molecules reaches the maximum possible density, allowed by their repulsive hard cores (Coulomb and Pauli repulsion). The hard cores at work, of course, are not those which correspond to the molecular ground state, but those produced by the state of the molecule fluctuating between the ground state and the excited configuration, the latter being usually more extended in space than the former. And it is just the molecular volume in the excited configuration that determines the density of the liquid.

The stability of the final configuration requires that matter and e.m. fields evolve in unison, their respective phases being locked in. This can happen only if the e.m. oscillations slow down; the common value ω' of this frequency of oscillation of the two intertwined fields (matter and electromagnetic) satisfies the inequality

$$\omega' < \omega \qquad (2)$$

whereas the wavelength remains constant. The constraint (2) guarantees that the e.m. field cannot be irradiated out of the system, and thus avoids the paradox that the state of minimum energy may go to one of still lower energy, which obviously does not exist.

So, at the end of the runaway, the collapsing system finds a stable dynamical configuration, where all molecules oscillate coherently in tune with an e.m. field trapped within the region (coherence domain, CD) whose diameter is the wavelength of the e.m. mode, whose frequency is given by eq.(1)

$$R(CD) = \lambda/2 = \pi / (E(2)-E(1)) \qquad (3)$$

Let us now extend this analysis from two-level molecules to multi-level molecules, such as water. In this case, a competition starts between the different e.m. modes, whose frequencies match the differences of energy of the different pairs of levels. The rate of growth of the e.m. field during the runaway towards the limit configuration depends exponentially on the critical value of density. Thus, in the case of many levels, the competition is gained by that level for which the critical density is the lowest; in a very short time the amplitude of the corresponding e.m. mode becomes so much larger than the amplitudes of the other modes, that all the available molecules are enrolled in this mode, leaving the other modes without resonating molecules and then suppressing the chance for their growth.

In the case of water, the selected excited level is a 5d configuration, whose energy lies 12.06 eV above the molecular ground state, so that the "e.m. field" coherence domain acquires a radius of 500 Angstroms. The corresponding critical density is just 0.31 g/cm^3 which coincides with the thermodynamical critical point. In Preparata (1995) it has been shown that molecules are at equilibrium with their field when they assemble themselves into an inner subsphere having a radius 3/4 of the CD radius, namely 375 Angstroms. That is the radius of the "matter" CD. The existence of other levels, apart from those involved in the coherent oscillation, produces however a peculiar feature. The oscillation between the two selected levels doesn't depend on the direct transition amplitude between them, but also on those ways of transition that occur when the system passes first through an intermediate level and then "leaks" into the final one. This complex structure of the transition occurs in water and, as shown in Arani et al. (1995) and Del Giudice (1994), gives to the e.m. field trapped inside the coherence domain the property to couple strongly with external systems possessing internal oscillation frequencies close to the frequency w' of the field. In the case of water, this frequency has been estimated 0.26 eV (in energy units) in Arani et al., quite close to the typical Debye temperature of many crystals. This suggests why water is such an effective solvent for many substances. We will come back to this remarkable property in the next section.

We introduce now the temperature in the present scheme. At $T>0$ the condensed system undergoes thermal fluctuations, produced by the collisions of the external particles or by absorption of e.m. radiation or other waves (sound or other). These assaults could give back to some molecules belonging to the coherent phase the energy spared in the above condensation. So, these particles are extracted from the coherent condensate and recover their freedom to become eventually part of the coherent domain again. On the other hand, they are prevented to escape from the system, since different coherence domains are compelled by the requirement to minimize their energy to coalesce together closing their ranks as much as possible. In Arani et al. it is shown that in such a case the water molecules in the outer region of a coherence domain are acted upon also by the evanescent tails of the fields of the neighbouring domains. This fact forces the fields of the neighbouring domains to lock in their phases, producing a unique e.m. field, coherent over a macroscopic region much larger than the single coherent domain.

The above feature squeezes the free spaces (interstices) between domains; on the other hand, molecules extracted by thermal fluctuations can move in those interstices only. The pressure they are subjected to prevents them to escape and leave the liquid. The fraction of extracted molecules, the "non-coherent" phase of the two-phase system, which is water, depends on T and increases with it. So, by increasing T, the coherent fraction decreases, populating the non-coherent fraction until the latter is able to break the squeezed regions of the coherence domains and the phenomenon of boiling occurs. In Arani et al. we have been able to calculate the boiling point of water.

It is interesting to consider the behaviour of the coherent fraction $F_c(T)$, as given by Fig.6 in Arani et al.: for $T<135$ K, F_c is about 1, a fact that can be understood by considering that the energy gap protecting the coherent state from thermal fluctuations is just 0.26 eV, a quite high barrier to overcome. In the above temperature interval almost all molecules belong to the coherent state, so that entropy is practically zero (in spite of the lack of spatial order). Since it is impossible to drag by a moderate shear force some molecules without moving all the others, viscosity has a very large value and consequently all the relaxation times are very long. For $T>135$K, $F_c(T)$, falls down by increasing its slope, until it reaches a maximum of the slope at 230K. The slope then decreases as $F_c(T)$ tends to zero at the critical temperature. It is interesting to know that in overcooled liquid water all the physical variables, such as the specific heat or the entropy, that depend on the rate of increase of the non-coherent phase, namely on $-dF_c/dT$, exhibit a maximum at about 230K.

When, by increasing T, the coherent fraction loses molecules towards the non coherent fraction, the volume of the extracted molecules shrinks, since they lose the contamination of the more extended excited configuration and revert to the less extended molecular ground state. Furthermore, the increase of T produces, as expected, the dilatation of the volume of the non-coherent phase. In conclusion, we get two opposite influences on the density of water as a function of T. The transfer of molecules from the coherent to the non-coherent phase produces an increase in density, whereas the dilatation of the non-coherent phase produces a decrease in density. At lower T, where $F_c(T)$ is larger, the first effect prevails, whereas at higher T, the non coherent fraction $F_n(T)=1- F_c(T)$ is dominant and the second effect takes the upper hand. So we can account for the density anomaly of water.

We conclude this section by showing how the well known tetrahedral coordination of water - i.e. each molecule in the more bound phase lies at the center of a tetrahedron whose vertices are occupied by its four nearest neighbours - emerges in the present framework. This result has been proved in Del Giudice et al. (1995b). The molecules of the coherent phase oscillate between the ground state and an excited 5d configuration. Since n=5, this configuration is quite extended in space, reaching a maximum distance from the molecular center of about 1.5 Å; being a d-state, the shape is quite prolate-like, so that the calculation of the electron charge distribution arising from the superposition of the two states gives a profile with two protuberances and two wells, the four features being regularly located on the surface along the tetrahedral directions. The hydrogen nuclei are just located in the direction of the wells. This restructuring of the electron cloud -which is the work of the electrodynamical process at the origin of the coherent state- now makes the electrostatic attraction between neighbouring molecules possible, phenomenologically described as "H-bonding". So, H-bonding is not actually the cause of the cohesion of liquid water, but the consequence of QED coherence. Once this has been recognized, all the phenomenological achievements of models based on H-bonding receive a dynamical justification in the present theory and can be fully recovered.

BIOLOGICAL PERSPECTIVES ARISING FROM QED COHERENCE IN WATER

The theory of water we have outlined so far opens some new ways towards an understanding of the dynamics of living systems. Actually, the main shortcoming of today's views, those generally accepted by molecular biologists, is a lack of understanding of the regulation of the traffic of molecules. Given the existence of selected chains of chemical reactions in living matter and the lack of chemical wastes, we must recognize that each molecule is actually able to meet only those molecules it should interact with in a meaningful way for the global interest of the being to which it belongs. How this seemingly theological course of events is casually determined by the physical dynamics of atoms and molecules appears mysterious and extra-mundane in a conceptual framework which includes only molecules moving freely and mutually interacting through collisions and very short range static forces. By which physical mechanism could these molecules, blind as moles, whose sensors to probe the environment are just the forces they are able to exhibit, and which can but explore a space not larger than two or three times their size, detect their prospective chemical partners at long distance, in the middle of a crowd of foreign and uninteresting other molecules? Of course, when the prospective partners reach each other, the close meeting of the chemical type occurs, following the prescription of molecular biology. However, should the meeting occur on a trial-and-error basis, since chemical bindings are not so exclusive and "monogamic", where are the outcomes of the unsuccessful trials, of the "biologically meaningless" reaction products, in other words the wastes that usually plague the chemical industries and are, on the contrary, wonderfully absent in the living chemical dynamics? The meetings of the ideal partners should be so rare that one, well acquainted with statistical physics, would expect that the simplest biological event might require a very long time.

The situation is completely different if these molecules are emerged in different medium, e.g. water, where a long range field, like the e.m. field, is able to transmit at long distances -hundreds of Angstroms, the size of a coherence domain, could be an appropriate scale- the e.m. messages coming from molecules. Given the peculiar structure of the coherent field in water, remarked in the last Section, if two molecules possess levels of about the same frequency in their spectra, which in turn should be close to the frequency of the coherent field in water (0.26 eV), then an effect of attraction between the two molecules arises, irrespective of how many other molecules, not resonating at the same frequency, might be in between. So, a system of recalls of different molecules at distance can be at work. Suppose now that an ensemble of similar monomer molecules gets assembled by the above mechanism (and also that the chemical short range forces can close the ties, forming a polymer), this new entity would then play the role of a coherence domain of monomers, which produces its own e.m. coherent field, with its proper frequency; this new field would bring together those molecules which can resonate at this frequency. The polymer would then act as a catalyzer of the chemical reaction of the co-resonating molecules. The newly formed molecules -in turn "coherence domains" of their monomers- would then catalyze on their own frequencies new reactions and so on. A non random scheme then emerges, where chemical and e.m. order are intermingled and each step of the e.m. sequence of formation of coherence domains of appropriate frequency fosters a further step in the chemical sequence, that in turn triggers a new step in the e.m. sequence.

Each step of the chemical sequence becomes, via the e.m. dynamics, the enzyme of the next chemical step. The specificity of each step is guaranteed by the co-resonance requirement that excludes non resonating strangers. The slowness of the usual collision mechanism of chemical reactions transforms itself now from a disgrace into a blessing, since it guarantees that the sequences occurring on the basis of the "frequency resonance recognition code" are not significantly subverted by unwanted random events. Furthermore, the energy output of the chemical reactions occurring at the boundaries of correlated regions, where energy can be transported in form of coherent excitations of a coherent medium and not in a diffusive way, would not spoil the order in the system, but could be the tool for the creation of new ordered structures.

The above scheme arises from the mere consideration of the potentialities of the theory based on QED, where matter is no longer a passive aggregate of ball-like molecules, but is an active medium able to select and catalyze the molecular reactions which may occur in its interior.

A concrete scheme giving a realistic description of the evolution of living systems has not been worked out yet, but this fascinating possibility begins now to appear as an intriguing perspective for a not too distant future.

MAGNETIC PROPERTIES OF THE COHERENCE DOMAINS OF WATER

The theory outlined in Section 4 shows that in those water molecules which belong to the coherent phase, the electron cloud oscillates in such a way that an electron is brought to a 5d configuration, close to the free electron region. The probability of quantum hopping into a nearby molecule is thus close to unity and, because of the electron phase coherence, the oscillating electrons of all molecules would concur in the same event. A hopping conduction could then be induced in the coherence domain, where the core formed by one electron per molecule would slide freely inside the coherence domain. The moment of inertia I of this rotation about an axis is:

$$I = (2/3)\,(4\pi/5)\,m_e(N/V)\,R^5 \qquad (4)$$

where m_e is the electron mass and R the radius of a CD. In the "natural unit system" I turns out to be $1.23\cdot10^5$cm. The spacing among the energy levels of this rotator is a multiple of the basic quantity

$$E_0 = I^{-1} = 4\cdot10^{-11}\ \text{eV} \qquad (5)$$

a very small quantity, corresponding to a frequency of about 35 kHz, which is close to frequencies measured in e.m. signals arising from water. Of course, a sphere made up of some millions electrons cannot be set in rotation by the random collisions of single molecules. The job could, however, be done by mesoscopic exchanges of energy, such as those provided by hydrodynamic turbulence. Once a rotation with angular momentum L and energy $E=L^2/2I$ is started, it would last indefinitely (the "effective temperature" for an ensemble of coherence domains with rotating electron cores is the absolute zero!), unless an hydrodynamic shock wave would come into play. So, in a motionless liquid, a rotational excitation of a coherence domain would have a very long lifetime.

Imagine now to switch on a magnetic field B: this is always the case on Earth, since the geomagnetic field is always there, with an intensity of half a gauss. The rotating electron core, which is endowed by the hopping current with a magnetic moment μ

$$\mu = (e/2m_e)\,L \qquad (6)$$

orientates itself along the direction of B. Since, as said before, the effective temperature $T_{eff} = 0$, then the Langevin formula guarantees that all cores are completely aligned with B. The total energy becomes:

$$E = L^2/2I - (e/2m_e)\, L\, B \tag{7}$$

The minimum of energy occurs when $dE/d|L|=0$, namely when:

$$L = (Ie) / (2m_e)\, B \tag{8}$$

Since the angular rotation ω is given by

$$L = I\omega \tag{9}$$

then

$$\omega = eB/(2m_e) \tag{10}$$

which yields $|\omega| = 1.5 \cdot 10^{-4}\,cm^{-1} = 4.6 \cdot 10^{6}$ rad/sec, which corresponds to a rotational frequency $\nu = \omega/(2\pi) = 0.7$ MHz.

The system will reach this equilibrium value if succussed and then allowed to dissipate any excitation energy it can have through hydrodynamic turbulence. We could furthermore introduce a much larger magnetic field B by a permanent magnet, producing much faster rotations of the electron cores of the coherence domains. This feature should be the dynamic agent of the bizarre phenomena reported at the end of Section 2.

CONCLUSION

Let us summarize the main conclusions of this article:

1. Every ensemble of molecules interacting through static forces only would necessarily become unstable once its density overcomes a critical threshold, which, for water, coincides with the thermodynamical critical density (0.31 gcm^{-3}). Thus, every theory of water which does not include electromagnetic interactions is unstable.

2. On the basis of the universally accepted rules of QED, a consistent theory of water can be formulated which accounts for the observed facts and is amenable to a two-fluid structure. There is a coherent phase, where all molecules oscillate in phase with a self-produced e.m. field of 0.26 eV or 2080 wave numbers and, doing so, restructure their own electron configuration considerably, making the tetrahedral

coordination possible. There is a non-coherent phase, which is a dense gas of the molecules extracted from the coherent phase by thermal fluctuations and increases with T, eventually producing boiling.

3. A fully autocatalytic scheme for biochemical reactions appears within reach by exploiting systematically the property of active medium that water receives by the existence of a coherent phase in its interior.

4. The magnetic properties of water, which have appeared magic so far, become for the first time ready for a scientific explanation.

References

Arani R, I. Bono, E. Del Giudice, G. Preparata, Int. J. Mod. Phys. B9 (1995) 1813

Bernal JD, R.H. Fowler, J. Chem. Phys. 1 (1933) 515

Buckingham AD in Neilson and Enderly (1986)

Dreyfus D, H.J. Wachmann, J. Chem. Phys. 76 (1981) 20

Del Giudice E, G. Preparata, Journal of Biological Physics 20 (1994) 105

Del Giudice E, G. Preparata, "Fast Sound in Water: Evidence for a Two-Fluid Structure ?" preprint MITH 95/11 of the Physics Dept. of the University of Milan, 1995a

Del Giudice E, A. Galimberti, L. Gamberale, G. Preparata, Mod. Phys. Lett. B9 (1995) 953, 1995b

Eiler Sundt: "Polar Magnetic Treatment of Water", published by the firm Olaf Fjeldsend A/S, P.O.Box 146, N-3201 Sandfjord, Norway

Franks F (ed.), Water - A comprehensive treatise, Vol.I, Plenum Press, New York, 1972

Neilson GW, J.E. Enderly (eds.), Water and Aqueous solutions, Adam Hilger, Bristol and Boston, 1986

Pastorino P, R.H. Tromp. M.A. Ricci, A.K. Soper, G.W. Neilson, Nature 366 (1993) 668

Piccardi G, La Ricerca Scientifica 29,6 (1959) 1 - in Italian

Preparata G, QED Coherence in Matter, World Scientific, Singapore, 1995

Roentgen WK, Ann. Phys. 45 (1892) 91

Schulte J, "Conservation of structure in aqueous ultra high dilutions", in the volume Ultra High Dilution", eds. P.C. Endler, J. Schulte, Kluwer Academic Publ., Dordrecht (1994), p.105

Sette F, G. Ruocco, M. Krisch, U. Bergmann, C. Masciovecchio, V. Mazzacurati, G. Signorelli, R. Verbeni, Phys. Rev. Lett. 75 (1995) 850

Soper AK, F. Bruni, M.A. Ricci, J. Chem. Phys. 106 (2997) 247

Stanley HE, J. Teixeira, J. Chem. Phys. 73 (1980) 3404

Stillinger FH, Science, 209 (1980) 451

Teixeira J, M.C. Bellisent-Funel, S.H. Chen, B. Dorner, Phys. Rev. Lett. 54 (1985) 2681

Walrafen GE, J. Chem. Phys. 48 (1968) 244

PATHOLOGY, COMPLEX SYSTEMS, AND RESONANCE

Paolo Bellavite and Andrea Signorini

A new vision of matter and life is emerging on the frontiers of science, particularly from the fields of quantum physics and mathematical theories and research which have yet to be systematized. Organisms are seen as highly regulated, complex, dynamic systems which display a characteristic meta-stability around certain homeostatic levels (Bellavite and Signorini, 1995). This meta-stability is the net result of continual oscillations, rhythms, networks, amplifications and feedback cycles. Living systems are "suspended" between order and chaos (Cramer, 1993; Bellavite et al., 1995); they partake of these two fundamental characteristics of matter and exploit them in a manner designed to promote survival.

Order and chaos are to be found at all levels of homeostasis, from the molecule to the human mind. We cannot see how these new perspective can fail to have an impact on the new orientations and trends in medicine. Medical theory, methodology and technology have always proceeded hand in hand with the general scientific theory and socio-economic situations of the times.

Living creatures are "open systems" far from equilibrium, subject to regulatory apparatus which cannot necessarily be represented by linear equations, and thus are capable of perceiving minimal perturbations, particularly when they are predisposed to such sensitivity, possibly by the very pathological process itself. New evidence from electromagnetism studies lends support to the possibility that living systems respond to extremely weak magnetic fields, and particularly to specific frequencies and specific intensity ranges. At the same time, studies regarding the physics of water suggest, or at least do not rule out, the possibility that water itself may act as a store and vehicle for electromagnetic oscillations.

J. Schulte and P.C. Endler (eds.),
Fundamental Research in Ultra High Dilutionand Homoeopathy, 105-116.

In this context, also the physiopathological changes leading to disease can be considered under a wider perspective and this new integrated concept of disease can not fail to have an impact on therapeutic approaches.

An integrated concept of disease

The behavior of any complex system, from single molecules to cells and to the whole body is dictated both by rigid structural or functional constraints and by other events or phenomena which are endowed with a higher degree of variability and freedom. These latter are represented by all the instances where the choice between two opposite or diverging states of the system is linked to minor and subtle variation of the conditions of the system itself or of its interactions with the environment. In far from equilibrium systems "bifurcation points" are found at each level of organization of living organisms, from molecules to cells, to spatio-temporal oscillations of physiological variables. Here we consider specifically these bifurcation points occurring in the dynamic progression of diseases.

The story and intimate nature of a pathological process presents various phases or aspects which integrate one another in sequences in time and space. If we are referring to the majority of diseases, which are not exclusively of genetic origin, what usually appears as the disease according to the traditional diagnostic criterion is the last phase, resulting in precise biochemical and anatomical abnormalities. Before this, however, there are at least three other phases: we have the first stage in which an initial disorder, mostly nonapparent apart from a number of very indistinct symptoms or variations in very subtle parameters, makes the body susceptible to perturbations induced by external agents. The subject cannot be defined as a sick person, but is more predisposed to falling sick than normal fully healthy individuals. In this stage we could, for instance, include those who are subject to overwork (stress) or to an unbalanced diet, those who smoke, who are exposed to low doses of ionizing radiation, or who present particular genetic characteristics making them statistically at risk (a number of HLA groups, etc.). To what extent this disorder is normal, in the sense that it is a simple reversible oscillation of a state of equilibrium, and to what extent it is pathological, in the sense that it generates disease in the presence of other perturbing factors, is an extremely subtle and hazy issue, so much so indeed that the same situations, which may even be quite severe, are supported as a normal burden of life by some people, whereas they are regarded as serious diseases by others (often defined as imaginary or psychosomatic illnesses). Clearly, at this level the balance between normal and pathological is extremely precarious, and the subsequent course of the disease can come down on side or the other according to shifts in minor factors.

A second bifurcation is to be found in the reactive phase of homeostatic biological systems. These systems - particularly the inflammatory and immune systems, but also the liver detoxification systems, the haemostatic system, and others - are "two-faced", i.e. they bring about healing, but they also cause damage because they can attack also the host. To what extent, in each individual case, the damage prevails over the restoration of the state of health or vice versa, depends on subtle variations in the behavior of the homeostatic system itself. In particular, the fate of the reaction depends on the "choice" that the system has to make between the price to pay, in terms of toxicity and suffering, and the guarantee of success of the operation in terms of survival of the body. For instance, in the presence of a lesion of the surface of a blood vessel, the haemostatic system comes into action to block the risk of haemorrhage and to initiate repair (clotting, platelet aggregation, increase in connective tissue and vascular musculature). Yet, through the same mechanisms a pathological event can occur: the haemostatic system entirely blocks the circulation of the blood in the vessel (thrombosis). What is it that tips the scale in favor of the positively-directed action rather than the unnecessary and frankly pathological one? It is the complexity of the multiplicity of mechanisms involved. A choice of this type depends, in fact, both on the individual elements involved (receptors, concentration of mediators, presence of exogenous chemical substances) and on the type of coordination available, on a "centralized" control system that assesses the information coming in from the various regions and elements involved and regulates accordingly the various responses. Thus, at the level of such a bifurcation, the outcome of the reaction may depend on an item of information that is significant in terms of the coordination of the reaction system or systems.

A third phase of the disease process, in which another very critical moment of decision presents itself, is when the reactive systems fail to cope with the situation and fail to rapidly restore the original state. At this point adaptation may set in, which is a semipermanent modification which, on the one hand, reduces the symptoms, but, on the other, may lead to various consequences, including deposition of toxins, excessive proliferation of cells, shifting of the receptor sensitivity threshold and biochemical and anatomical changes that defer the problem for lengthy periods or shift the pathological consequences from one organ to another. Adaptation makes it possible to live with the disease, but also constitutes, in a certain sense, a renunciation of complete healing. Also the choice between reaction and adaptation is very complex, being determined by a number of factors.

From a biophysical standpoint, disease may be regarded not only as a functional or molecular-structural abnormality, as in the classic view, but also (and not by way of contrast) as a disturbance of an entire network of electromagnetic communications based on long-range interactions between elements (molecules, nerve centers, organs, to mention but a few) which oscillate at frequencies which are coherent and specific and thus capable of resonance. This would be a disturbance of internal oscillators and their communications. Our knowledge is still too scanty to say whether or not these oscillators can be identified with certain nerve centers in particular (the ability to oscillate at characteristic frequencies is typical, though not exclusively so, of nerve centers) or with the collective behavior of nerve centers and/or other tissues or cells. With regard to cell to cell and intracellular communication, biophoton research suggests the existence of extensive coherent fields interaction in the visible range (Ho and Popp, 1993).

Chaos and fractals

In dynamic system that are far from thermal equilibrium, minimal variations in the conditions of the system (such as those induced even by a very small oscillatory resonance) may play a decisive role in the subsequent evolution of the system itself. The new concepts emerging from chaos and fractal studies can help to interpret and possibly dominate these complex events. In a variety of systems, the butterfly effect may be used to control chaos, on condition that the parameters to be controlled and changed are well known (see Shinbrot et al., 1993; Schiff et al., 1994; Moss, 1994). The relationship between these new mathematical theories (and experiences) and homoeopathy can hardly fail to be extremely close both because the communication of information from the solute to the solvent and from highly diluted compounds to the disordered (ill) organism is nothing other than the transition from chaos to order, and because the dilution/dynamization processes has been suggested to follow fractal dynamics observations (Garner and Hock, 1991; Shepperd, 1994). The term fractal was coined in 1975 by B.B. Mandelbrot and gained extensive notoriety in scientific circles in the early '80s (Mandelbrot, 1982). What is meant by this term are those mathematical or geometrical entities which are endowed with a fractional dimension (from the Latin fractus, meaning "broken"). Many fractal figures have a repetitive configuration on changing scale, a sort of self-similarity between details and the general pattern.

Fractal shapes can be generated by the computer using algorithms (lists of instructions which specify the operations to be performed to solve a given problem) using mathematical functions which are suitably iterated (i.e. repeated for the desired number of times, each time using the result of the previous calculation as the basis for the following). By means of these operations, two- or three-dimensional figures with the following characteristics appear: a) an enormous variety of details of different shapes,

b) the presence of subtle ramifications that can be pursued in the finest detail, c) self-similarity, whereby, on magnifying part of the structure, details can be detected which repeat themselves on different scales of magnification.

In very complex mathematical structures such as the sets of Julia and Mandelbrot, extremely varied and fanciful details can be observed (circles, spirals, helixes, stars, various ramifications), within which other, different details can be discerned on magnifying the image. Within some of these particular images one finds, surprisingly, mini-sets very similar to the macroscopic ones from which they originated. In the close-up detail we rediscover an image which appeared to have been lost in the variety of details and ramifications. By increasing the number of iterations, a better definition of the fractal image is obtained.

Fractal patterns are easily found in chaotic systems, representing elements of regularity. In several chaotic functions, on continuing the iterations and further increasing the coefficient value, after the periods of chaos periods of order may reappear, followed by new zones of chaos and then order. There is thus a recurrent regularity (Hofstadter, 1991) in successive generations of transitions from chaos to order, with the reappearance of single solutions or regular oscillations which undergo cascade duplication on increasing the coefficient value. This recurrent regularity creates figures with regularity and irregularity "bands" which are repeated and resemble one another, with a fractal type pattern.

Mandelbrot and, more recently, other investigators have observed that many apparently disorderly natural objects possess this fractal property. This is having a very considerable impact in the scientific world. Despite the fact that these forms were discovered by a mathematician and are still studied mainly by mathematicians and computer scientists, fractals are useful instruments for describing a whole variety of physical phenomena and natural forms. One example of a fractal form is a tree, whose trunk is divided into branches; the branches themselves then divide into smaller branches, twigs, and so on until you get to the leaves, which in turn present veins with multiple subdivisions. Other examples of natural fractals are clearly illustrated by certain flowers and snowflakes, as well as by non-crystalline molecular aggregates, viscous ramifications in immiscible fluids, corals, electrical discharges such as lightning, the ramifications of the airways and blood vessels, the dendrites of the neurones, the Purkinje system conducting electrical signals in the heart, and the folds of the intestinal mucosa. It has also been demonstrated that, in many different physical situations, particles floating on the surface of an irregularly moving fluid display a fractal arrangement (Sommerer and Ott, 1993). Fractal geometry thus refers to some form of conditioned randomness, so much so, indeed, that some people talk about a determinism of chaos (deterministic chaos).

Various experiments have suggested that the biological activity of homoeopathic dilutions does not diminish or increase regularly with increasing dilutions, but follows a "pseudo-sinusoidal" trend, with peaks of activity and troughs of inactivity. It has been found that the periodic recurrence pattern of pseudo-sinusoidal activity is not regular, but chaotic and unpredictable, which implies the possibility that an effective information re-presents itself after a number of log dilutions. According to the logic of current chemical reasoning, such a trend appears thoroughly and quite definitely absurd, but in the light of chaotic phenomena and the nonlinearity of many biological mechanisms perhaps some kind of underlying logic can be traced.

If we are prepared to give credence to such paradoxical results (note that most of the reported experimental data are still awaiting unequivocal confirmation by independent laboratories), one question inevitably springs to mind: how can information disappear and then reappear? Where did the information go in the mean time, while the inactive dilutions were being done, between one peak and another? To answer this question the fractal "line of reasoning" may come in handy. In fact, if we are to take the results of the experiments seriously, we have to admit that the information of the compound dissolved is not completely "dissipated" in the course of the successive dilutions, even when the dilutions are inactive. Evidently there should exists some mechanism of transmission and storage of the information in the course of the dilutions such that the next dilution following an active dilution may give rise to a form (or vibrational frequency) differing from the previous one (and for this reason inactive), but capable, after a few steps of further diversification, of causing the original (active) information to reappear. The dilutions, then, would not produce a loss of information (increased entropy), but only a change and variety of forms, which in turn may regenerate the starting form. Such behavior is reminiscent of what we have seen apropos of the mathematical iterations that generate fractals. This, then, would be a phenomenon similar to the "recurrent regularity" typical of chaotic systems.

The hypothesis has been advanced (Gardner and Hock, 1991) that the successive dilutions and dynamizations performed in the preparation of a homoeopathic remedy introduce an element of information gain, as is observed in the Mandelbrot sets with successive iterations. It is suggested that low dilutions (few iterations) produce poor definition of details and carry rough and imprecise information, whereas high dilutions (many iterations) are characterized by better definition of details, as can be seen in the profiles of the Mandelbrot sets. If the dilutions/iterations are few, the image is blurred, whereas, if they are repeated many times, the image is precise and, surprisingly, "reappears", i.e. it is reproduced in detail in subsets and in subsets of subsets. The image of a certain structure (in the case of homoeopathy, the mother tincture) reappears in a "similar" form in successive dilutions, practically to infinity.

Such a phenomenon may basically be responsible for the fact that in classic homoeopathy the high dilutions are regarded as more specific and profound in their therapeutic effect if there is perfect matching of the symptoms of the patient and the remedy, i.e. if the details" of the analogy have emerged clearly from the homoeopathic history-taking. In practice, the fewer the symptoms shared by the patient and the remedy, the lower will be the dilutions used; the more symptoms they have in common, the higher will be the dilutions prescribed.

We must acknowledge that this discussion is highly speculative and hypothetical. Our aim in referring to these aspects is to stress how a scientific approach to homoeopathy in future will require the joint contribution of several disciplines, including mathematics, geometry and computer sciences. In general terms, the suggestions we are making here emphasize the fact that those researchers who are prepared to engage in the study of the unsolved, complex problems of biology and physics (including the study of homoeopathic high dilutions) must begin to include the dimensions of chaos and the fractals in their conceptual, and possibly also in their experimental armamentarium.

Resonance and the homoeopathic approach

We shall attempt here to summarize a few points with the aim to present a frame of reference for putting homoeopathy into rational and experimentally viable perspective. Of course, the following points should be underlined: a) hypotheses are essential to the evolution of knowledge, but one has to guard against presenting them as certainties, b) the following discussion is based on the assumption that highly diluted remedies prepared according to the homoeopathic methods are endowed with specific information of biophysical nature (quasi-particles, water clusters, isotopic lattices, and so on), a highly controversial issue that is discussed in other sections of this book, c) here we restrict our consideration to the possible mechanism of interaction between high-dilution homoeopathic remedies and the organism, that is only one of the various questions raised by the homoeopathic approach and d) a biophysical perspective does not exclude that many effects of low- and ultra-low doses of drugs are due to conventional molecular interactions. The hypotheses set forth here also refer to theories proposed by other authors (Vithoulkas, 1980; Tetau, 1985; Callinan, 1986; Rubik, 1990; Popp, 1990; Ullman, 1991; Schulte and Endler, 1994; Poitevin, 1995).

If the disease process involves a disturbance of oscillation frequencies - and of the communications associated with them - it should be brought back to a state of equilibrium by means of synchronization or tuning, i.e. by means of a change in frequency imposed by interaction with another oscillator. According to this notion, the homoeopathic remedy might act in the patient as an external guide frequency.

The phenomenon of resonance is well known in physics, where it occurs in many fields: acoustics, mechanics, and electromagnetism, as well as nuclear physics. By virtue of this phenomenon, a system which is characterized by its own oscillation frequency can enter into vibration if stimulated (subjected to sound waves, electromagnetic waves, or mechanical vibrations according to the nature of the system) by frequencies close to those peculiar to the system itself. If the system is already oscillating, the resonance may greatly increase the amplitude of the oscillation, whenever the waves overlap, while the opposite may also occur, namely an arrest of oscillation, if the interaction is between two waves of the same frequency but opposite in phase. Of course, biological systems are characterized by a very complex set of frequencies, inkeeping with the complexity of their components. For resonance phenomena to occur, the frequencies do not need to overlap exactly; it is only of one or more harmonics (harmonics are the simplest components into which periodic functions produced by their overlap can be broken down) match; the harmonics of a given periodic system all have frequencies which are multiples of the fundamental frequency, called the first harmonic.

Resonance, then, is a way whereby information is transmitted between two similar systems (as regards vibrational or harmonic frequencies) without structural modifications and without the passage of matter. These linkage phenomena between oscillators, which generate synchronism and cooperativity, are of paramount importance in many physiological functions, particularly in the nervous system, but also in the cells regulating cardiac rhythm, in the cells secreting insulin in the pancreas, in the ciliated epithelia, and in the involuntary contractions of smooth muscle (Breithaupt, 1988; Engel et al., 1992; Strogatz and Stewart, 1993). Interestingly, it has been suggested (Tsong, 1992) that resonance may transduce information between an oscillating, low-level, electromagnetic field and, molecular sub-domains of cell enzymes. This type of interaction should lower the activation barrier of the rate-limiting step of enzymic reactions, thus increasing the overall catalytic activity.

On this basis, a hypothetical model of the possible action mechanism of homoeopathic drugs can be advanced. A potentized homoeopathic drug might be regarded as a small amount of matter containing elements oscillating in phase (coherently), capable of transmitting these oscillatory frequencies, via a process of resonance, to biological fluids (in turn mostly made up of water), but also to complex "metastable" structures, subject to nonlinear behavior patterns and capable in turn of oscillating (macromolecules, a-helixes, membranes, filamentous structures, receptors). There would thus be the possibility of a link between drug frequencies and oscillators present in the living organism perturbed by the disease.

Even signals which are extremely small, but which are endowed with highly specific information and are capable of resounding in unison with the recipient system, could act as regulators, if it is admitted that the dys-regulated system or systems are in a state of precarious equilibrium, near to the bifurcation" point, where the choice whether to move in one direction or the other is related to minimal fluctuations on the border between order and chaos.

Homoeopathy might therefore act on the initial "decision-taking" levels of the body's repair and defence systems. When a disease reaches the stage where gross biochemical and anatomical consequences of the disease process are present, we are entering a field in which there would appear to be a much greater indication for the use of strong therapies such as surgery, replacement therapy, or the use of drugs at high doses, though even here a possible contribution of homoeopathy should not be ruled out (provided at least some of the homeostatic control systems can intervene).

Even as referring to the diagnostic sphere, it is clear that the more a disease claims attention in terms of biochemical and anatomical abnormalities, logically the greater will be the tendency to resort to laboratory investigations and diagnostic imaging techniques, whereas a homoeopathic diagnostic work-up aimed at capturing the subtle differences in personality and symptoms between one patient and another would make very little sense. Conversely, however, conventional diagnostic means can achieve very little within the framework of the initial subtle changes in complex homeostatic equilibria, or, even if they manage to pinpoint individual variations in biochemical or functional parameters, yield no criteria for reconstructing the picture as a whole and thus for implementing a complete therapy. Depending upon the level at which they operate, different methodologies of both diagnostic and therapeutic type are adopted. Homoeopathy is a probe into complex systems. Despite all the limitations related to difficulties in rendering the homoeopathic approach objective, it is clear that it entails an attempt to "explore" the patient's medical history at the level of the neuro-endocrine system and thus to calibrate some form of therapeutic intervention at this level, too. Homoeopathy and toxicology regard inflammation as a symptom (i.e. as a signal or message) and not as a disease, and they regard this symptom as the expression of an alteration of the relationship between subject and environment and/or between systems in the same subject. In the light of what we have said about the complexity of living systems, these concepts appear to be of great topical interest, quite apart from the difficulties encountered in rationalizing and perhaps even in demonstrating everything that homoeopathy claims.

Homoeopathy used with ultra-diluted drugs is thus a tentative approach to the bioenergetic regulation of the human body, utilizing a physical-biochemical interface due to the extreme sensitivity of biological systems to this type of regulation. The potential strength of the method consists in the fact that it attempts to achieve the maximum possible degree of specificity of the exogenous regulatory intervention. As stated earlier, the effective doses will be lower, the more specific the stimulus and the more sensitive the target system. If we admit that information is contained in metamolecular form in the homoeopathic remedy, this information may also act in a metamolecular manner in the bioenergetic target system.

How can the maximum specificity of information be achieved, if we know so little about such bioenergetic systems? The answer is in the main principle of homoeopathic tradition, the "law of similars". This fundamental principle, based, as it is, essentially on the observation of effects (i.e. on comparison of the effects of the drug with those of the disease), is in a certain sense independent of any knowledge of the mechanism which causes the effects and thus also applies to the metamolecular level, once we have admitted the existence of the latter. Further details regarding this important point of homoeopathic theory can be found in the cited work of the Authors (Bellavite and Signorini, 1995).

Homoeopathic reasoning is based more on analogy than on inductive thinking. The use of analogy (i.e., identification of similarity) is justified on theoretical grounds on the basis of the fact that the various elements of reality are interlinked, because they all derive from the same evolutionary process; in nature, we find the result of a growth of items of information which are always kept in contact with one another. Animals have always lived in contact with vegetable substances and minerals, and it is for this reason that a molecule contained in a flower may be "similar" to molecules contained in the animal, and there may therefore be a transfer of information. Information is transferred only between similars, or between opposites, or in any event between elements that are capable of interacting as a result of affinity of structure or of vibrational frequencies (harmony, resonance, coherence). Analogical reasoning consists in grasping this basic principle.

The "secret" of homoeopathy lies in the meticulous gathering of information both in the proving phase and in the homoeopathic history-taking phase. This information can come from the hidden depths of the homeostatic regulatory system under investigation, but is still information. In the homoeopathic method it is used directly in the therapeutic intervention, trusting in the fact that the body will know how to receive, decode and utilize this informational input for the purposes of restoring the lost equilibrium.

Another "secret" of homoeopathy is that it deals with the human being as a whole, devoting the maximum attention to symptoms of a psychological type and those peculiar to each individual subject (individualization). In this way, it achieves a very substantial measure of specificity, inasmuch as it is now well known that the response to drugs can vary on the basis of the characteristics of the individual user.

Finally, we have to mention the fact that we are witnessing a convergence of the homoeopathic and the acupuncture approaches. A series of controlled studies on the conductance of acupuncture points suggests (Kroy, 1989; van Wijk and Wiegant, 1994; see also Bellavite and Signorini, 1995), that the sensitivity of a given patient to a given compound (either toxic or beneficial) can be detected by measuring the disturbances of electrical currents on the acupuncture points. If this will prove true and reproducible, we would thus have another key to penetrating the sanctuary of biological and pathological information. Gathering information, such as knowing, for instance, whether a certain patient is allergic to or intolerant of a certain compound, or knowing whether a certain drug restores the balance or produces electrical dys-equilibrium would be an undeniable step forward from the point of view of diagnosis and therapy.

An interpretation of homoeopathy such as that presented here reconciles the integrated view, which considers the complexity of the human being in all his or her components, with the reductionist view, which considers the single organ, cell, or molecule. There cannot be a contrast between the whole and the fragment which this whole contains and therefore various medical approaches should be utilized according to which level of integration and which physiopathological mechanism(s) is the object of treatment. In fact, earlier in this volume Ho pointed out that an organized factorization on a cellular level may account for the actual functionality of an organism as a whole.

References

Bellavite, P., Signorini, A. (1995) *Homoeopathy, A Frontier In Medical Science. Controlled Studies And Theoretical Foundations.* North Atlantic Books, Berkeley. Italian edition: IPSA Ed., Palermo

Bellavite, P., Andrighetto, G.C., Zatti, M. (1995) *Omeostasi, Complessità e Caos, Un'INtroduzione.* Franco Angeli, Milano. Boiron, J., Belon, P. (1990) Contributions of fundamental research in homoeopathy. *Berlin J. Res. Homoeopathy* 1: 34.

Boiron, J., Belon, P. (1990) Contributions of fundamental research in homoeopathy. *Berlin J. Res. Homoeopathy* 1: 34.

Breithaupt, H. (1988) Biological rhytms and communication. In: *Electromagnetic bio-information* (Popp, F.A. *et al.*, eds.). Urban & Schwarzenberg, Munchen, p. 18.

Callinan, P. (1986) L'énergie vibratoire et l'homme. Un modéle pour le mode d'action de l'homéopathie. *Homéopathie francaise* 74: 355.

Cramer, F. (1993) *Chaos and Order. The Complex Structure of Living Systems.* VCH Verlagsgesellschaft, Weinheim (Ger.)

Davenas, E., Beauvais, F., Amara, J., Robinson, M., Miadonna, A., Tedeschi, A., Pomeranz, B., Fortner, P., Belon, P., Sainte-Laudy, J., Poitevin, B., Benveniste, J. (1988) Human basophil degranulation triggered by very dilute antiserum against IgE. *Nature* 333: 816.

Engel, A.K., Konig, P., Schillen, T.B. (1992) Why does the cortex oscillate? *Curr. Biol.* 2: 332.

Garner, C., Hock, N. (1991) Chaos theory and homoeopathy. *Berlin J. Res. Homeopathy* 1: 236.

Ho, M.W., Popp, F.A. (1993) Biological organization, coherence, and light emission. In: *Thinking About Biology* (W. Stein and F.J. Varela, eds.). SFI Studies in the ciences of Complexity, Lect. Note vol. III, Addison-Weisley Pub. Comp., Reading, MA, p. 183.

Hofstadter, D.R. (1991) Attrattori strani: enti fra ordine e caos. In: *Il Caos. Le Leggi del Disordine* (G. Casati, ed.) Le Scienze S.p.A., Milano, p. 71.

Kroy, W. (1989). The use of optical radiation for stimulation therapy. In: *Electromagnetic Bio-Information* (Popp, F.A. *et al.*, eds.). Urban & Schwarzenberg, Munchen, p. 200.

Mandelbrot, B.B. (1982) *The Fractal Geometry of Nature.* W.H. Freeman & Co., New York.

Moss, F. (1994) Chaos under control. *Nature* 370: 596

Poitevin, B., Davenas, E., Benveniste, J. (1988) In vitro immunological degranulation of human basophils is modulated by Lung histamine and Apis mellifica. *Brit. J. Clin. Pharmacol.* 25: 439.

Poitevin, B. (1995) Mechanism of action of homeopathic medicines. Recent findings and hypotheses 1: Physicochemical mechanisms. *Brit. Hom. J.* 84: 32

Popp, F.A. (1990) Some elements of homoeopathy. *Brit. Hom. J.* 79: 161.

Rubik, B. (1990) Homeopathy and coherent excitation in living systems. *Berlin J. Res. Homeopathy* 1: 24.

Sainte Laudy, J., Belon, P. (1993) Inhibition of human basophil activation by high dilutions of histamine. *Agents and Actions* 38: C245.

Schiff, S.J., Jerger, K., Duong, D.H., Chang, T., Spano, M.L., Ditto, W.L. (1994) Controlling chaos in the brain. *Nature* 370: 615.

Shepperd, J. (1994) Chaos theory: implications for homeopathy. *J. Am. Inst. Hom.* 87: 22.

Shinbrot, T., Grebogi, C., Ott, E., Yorke, J.A. (1993) Using small perturbations to control chaos. *Nature* 363: 411.

Schulte, J., Endler, P.C. (1994) Preliminary elements of a theory on ultra high dilutions. In: *Ultra High Dilution* (P.C. Endler and J. Schulte, eds.). Kluwer Acad. Publ., Dordrecht, p. 245.

Sommerer, J.C., Ott, E. (1993) Particles floating on a moving fluid: a dynamically comprehensible physical fractal. *Science* 259: 335.

Strogatz, S.H., Stewart, I. (1993) Coupled oscillators and biological synchronization. *Sci. Am.* 269 (3):1993

Tetau, M. (1985) Recepteurs et hom_opathie. *Cah. Biother.* 88: 29.

Tsong, T.Y. (1992) Molecular recognition and processing of periodic signals in cells: study of activation of membrane ATPases by alternating electric fields. *Biochim. Biophys. Acta* 1113: 53.

Ullman, D. (1991) *Discovering Homeopathy: Medicine for the 21th Century.* North Atlantic Books, Berkeley.

Vithoulkas, G. (1980) *The Science of Homeopathy.* Grove Press Inc., N. York.

INTERACTIVITY, FEEDBACK AND CHAOS CONTROL

K. W. Kratky

In recent years, a shift in emphasis in several scientific fields could be observed from 'understanding' and 'predicting' to 'action' and 'control'. One such field is chaos research, where the loss of long-term predictability is accompanied by the possibility of controlling such systems - even without knowledge of the underlying rules or equations. There are also systemic therapy forms. In linguistic terms this means the transition from a semantic to a pragmatic viewpoint. From an epistemological standpoint, it means the change from a passive to an (inter-)active observer. But it goes further: terms such as action or control change their own meaning through the introduction of feedback loops. Thus the chaos control in question is in fact more of a process of chaos regulation, comparable to many processes in the human organism. Systemic therapy, on the other hand, corresponds more to self-healing than to external intervention. Finally, chaos control using feedback, which is being looked at in more detail here, is compared to various "feedback therapies" (such as bio-resonance therapy). Lastly, a possible connection with homoeopathy is considered.

In the classical perspective, (natural) scientists observe the world from the "outside", thus developing models and, eventually, theories. These are tested by comparing predictions with experimental results. With the availability of workable models and theories, it becomes possible to control nature to a certain extent. However, this is no longer the domain of scientists, but that of technicians and engineers.

Separating observation from intervention neglects the feedback loop of the total process. But this is no exception. The deeper reason is to be found in science's claim to be able to understand the world at the end of the day. In order to tackle this "great project", the world needs to be split up into parts. In this way, nature becomes detached not only from the scientists, but also from itself, as for instance

J. Schulte and P.C. Endler (eds.),
Fundamental Research in Ultra High Dilutionand Homoeopathy, 117-125.

in the context of a reproducible experiment. On the other hand, it is just in that context that strong connections are simply assumed, while in reality they may be rather loose: e.g. between analysis and synthesis, problem and solution, diagnosis and therapy. Physicists are penetrating deeper and deeper into the micro world in the hope of being able to return to our macro world. In medicine, on the other hand, enormous efforts are being made in order to improve diagnostic methods. But is this making therapy easier and better? In all these cases we are spending a lot of time and money on the first step and often fail to reach the second.

A systemic or cybernetic viewpoint may help to shed some light on this problem. What changes are brought about by an appropriate consideration of feedback? Science's detachment from the world can no longer be maintained. Some form of constructive viewpoint needs to replace the spectator perspective. An intuitive feeling for which interactions are important takes the place of conventional science. This is more of an artistic, or technical, approach. However, science itself is changing, especially in the areas of chaos and system research (Kratky 1991, 1992).

Chaotic systems are characterized by the impossibility of making long-term predictions. Feedback loops and corresponding non-linear interactions (see below) lead to a sensibility towards all kinds of small influences, which suggests an integral view of these systems. This means that the classic, detached point of view can barely be maintained. On the other hand it turns out that chaotic systems can be controlled with astonishing ease (while, until a few years ago, the opposite was believed). They are not only sensitive to the slightest unwelcome disturbance, but also towards small, welcome influences. This leads to two possibilities. Either the conventional perspective is maintained of the passive scientist, who is removed from the object of his or her observation, and who regrets the loss of predictability. Or the opportunity to control systems in a new way, which requires an (inter)active viewpoint, is welcomed. This may be compared to a high-rope walker or surfer, who constantly has to balance the movements with small corrections, without knowing in detail how that balance act is actually accomplished.

The control of complex systems

Following on from the above theoretical considerations, we now turn to a mathematical perspective, which is kept as simple as possible. The behavior of systems is usually modeled by a set of coupled differential equations. In simple cases, three equations with the variables x, y and z are sufficient:

$$\begin{aligned}\dot{x}(t) &= f(x(t),y(t),z(t)) \\ \dot{y}(t) &= g(x(t),y(t),z(t)) \\ \dot{z}(t) &= h(x(t),y(t),z(t))\end{aligned}$$
(1)

Here, f, g and h are functions of these three variables. The dot on top $\dot{x}(t)$ denominates the change of x in the course of time, $\dot{y}(t)$ and $\dot{z}(t)$ accordingly. The three equations, or rather their solutions, thus describe the system's dynamics. The actual meaning of x, y and z depends on the respective problem at issue. Let us assume that variable x is sensitive to observation and to perturbation. The system in question may be for instance an electronic switching circuit, where electricity or voltage are measured with within the circuit.

In the following discussion we are interested in complex dynamic systems. These are characterized by the fact that they may exhibit varied and surprising characteristics, even if their constituent equation of motion appears to be as simple as the simultaneous system of equation (1). In order for complex dynamics to evolve, it is necessary that at least one of the constituent equations is non-linear (Mahnke et al. 1992). A typical linear function is:

$$f(x(t),y(t),z(t)) = ax + bx + cx \tag{2}$$

where a, b and c are constants. A non-linear function would contain such terms as x^2, yz or $\sin(y)$.

In a typical system like (1), depending on the initial values x_o, y_o, z_o at time t=0, an internal dynamic can develop, and if the system of equations describes (or model) a physical system, the variable x, y and z may be experimentally accessible. As the system propagates in time, characteristic patterns of motion, so called attractors, will usually develop. A characteristic dynamic of a system is demonstrated at the well known pendulum. The pendulum has two points where it can rest. One is at the very bottom of its motion which is called the stable equilibrium (a simple *attractor*). The other one is at the very top when it is turned over. A very small perturbation at this position and the pendulum swings towards its stable equilibrium (it is *attracted* towards the stable equilibrium). Hence, the top resting point is called unstable equilibrium. In terms of complex system language, the two types of equilibrium positions of the pendulum are called stable and *unstable fixed* points, respectively. Similarly to a balance act, the unstable equilibrium may be maintained, or controlled by way of small, well attuned corrections.

In complex systems there exist two further types of characteristic dynamics, periodicity (cycle, oscillation) and non-periodicity. The chaotic type of periodicity, however, cannot be predicted on a long term basis, hence it is called *chaos*. In some cases dynamic periodicity may appear to be mere coincidence. In order to distinguish mere coincidence from this type of chaotic behavior, chaos often called *deterministic chaos*. As in the case of fixed points, periodic and non-periodic dynamics have unstable (hidden) and stable (manifest) variants, so called periodic and chaotic attractors.

Controlling (dynamic) chaotic behavior generally means control of the chaotic dynamics in the sense that one of the unstable periodic orbits "hidden" in a chaotic attractor is stabilized at the expense of the latter. Various methods have been developed to control chaotic dynamics (Hübinger 1993, Pyragas 1992). Here we follow two versions introduced by Pyragas, the external force control and the delayed feedback control, respectively. External force control may be considered the more conventional course of action: necessary influences are determined which, when created from the outside, stabilize a desired dynamic (which could be interpreted as being "healthy"): e.g. the cycle $x^*(t), y^*(t), z^*(t)$. In order to control the dynamic of the system, only one of the three constituent equations in (1)) needs to be subject of external forces, e.g

$$\dot{x}(t) \quad = \quad f(x(t),y(t),z(t)) + K\left[x^*(t)-x(t)\right]. \tag{3}$$

Here it is obvious that a desired dynamic which includes the external control has to have features such that the simultaneous system of equations (1) yields the same solution. K is then a parameter which can be adjusted in such a way that it enables optimal ("desired dynamics stabilizing") control. It corresponds to the strength of correction, which should neither be too weak, nor too strong (as the high-rope walker is well aware of). Pyragas' second method has more to do with regulation than control. Though this method of chaos control appears to be very simple, it proved to work astonishingly well. One of the appealing features of this method is that an exact knowledge of about the system is not all required - in a sense it is sufficient to "reintroduce" only one observable, say $x(t)$, into the system at a certain time delay (introduction of additional feedback). The first equation of our model description of the dynamical system looks then

$$\dot{x}(t) \;=\; f(x(t),y(t),z(t)) + K[x(t-\tau)-x(t)]\,, \tag{4}$$

with τ being a time delay parameter enabling us to stabilize cycles with periods of length τ. This method even works in case of a large mismatch of cycle lengths. Naturally, the closer the dynamic approaches the desired cycle, the smaller this additional term becomes, until it only oscillates slightly around some small value. A more detailed description of this method can be found in Kratky (1995, 1996).

For this type of chaos control, it has to be noted that the dynamics is not controlled by an purely "external" force. Instead, an already latently present dynamic of the system is extracted and fed back into the system. In this way, small controlling events are sufficient for both the first (3) and the second (4) method. The difference between the methods lies in the fact that the external contribution is dominant in the first, while in the second the internal processes play the main role. Of course, it always possible to significantly change the original system using strong effective additional forces, thus forcing practically any desired dynamic onto it. However, the methods introduced above are more intelligent, and also far more energy efficient, since they work *with* the system, and not against it.

Discussion

As mentioned above, the term "control" changes its meaning in the presence of feedback techniques, namely from a directed interpretation to self-control, or rather regulation. The human body, too, controls itself, using a large number of feedback loops. For this, very small influences, which do not use much energy, are sufficient. Therefore, chaos and its regulation are no accident of nature. Rather, they have turned out to be optimal in the course of evolution. This aspect is also becoming more and more important for our civilisation. Accordingly, in the near future many uses are going to be found for chaos control in industry and technology. However, to many engineers this currently still appears too unsafe and too dangerous. They prefer to stick to streets and railroad tracks, instead of "surfing" (as mentioned above).

The human organism's ability to control itself effectively is strongly related to the existence of a dynamic healthy state. Medical treatment in the case of illness (e.g. prescribing drugs) may be interpreted as changing the organism's parameters or variables. There are complementary medical methods which close the loop between diagnosis and therapy such as in the case of eg. bio-feedback which enables a person to have control of over the heart beat, blood pressure etc., a method mostly used for the purpose of health maintenance, rather than in the treatment of illness. To the person involved, this control of variables, which are normally out of reach, will seem fairly mysterious, particularly since she or he does not really how such control is actually achieved.

Chaos research is increasingly gaining in importance. It not only causes an expansion of existing knowledge, but also seems to be leading to a new weltanschauung. Thus, loss of predictability, coupled with a gain in controllability, are not compatible with the conventional view. Chaos and system research could of course be excluded from the sciences. However, since this is obviously not happening, the image of the scientist may change from that of a "passive" one to that of an "interactive" observer. To act will no longer mean that one has to know everything beforehand. Rather, it will require sensitivity, and the ability to react immediately to the environment's reaction. This may also have a profound impact on areas such economy, ecology and medicine.

Interestingly, in traditional chaos control Pyragas' approach (1992) is less popular than other methods, such as eg. the one proposed by Ott et al. (1990). Is this so, precisely because Pyragas' chaos control is simple and works, while a deeper understanding of why it works is still missing? This leads one to the comparison between orthodox and complementary medicine, while remembering that medicine is a special case, since it was never (merely) a science, but always an art and a trade as well. Nevertheless, therapies which, though effective, are not fully understood, are regarded with suspicion. Chaos research may change this by increasingly shedding light on the organism's regulation, showing that healing is first and foremost self-healing. In this way the gap between "scientific" and "unscientific" medicine may soon close.

This problematic can also be viewed from the perspective of linguistics or information theory, which necessitates a differentiation between syntactic, semantic and pragmatic information, as the following example shows. When speaking to someone, all sentences should be formally (grammatically) correct as well as having a certain content and an understandable meaning. This relates to the syntactic and semantic aspects. In addition, there will also be an intention behind the use of these particular sentences. A speaker may try to influence, or even manipulate, listener. That would be a pragmatic aspect of communication.

Syntactic, semantics and pragmatism usually, but not necessarily, belong together. Thus a formally correct sentence does not necessarily have to make sense. In a sense we are living in three worlds. When working with computers, the formal aspect is very important, while the semantic aspect dominates the conventional sciences. In personal relationships as well as in the system sciences, the pragmatic aspect is central. Thus, systemic family therapy tries to heal through communication, and the creation of additional feedback loops. The therapist concentrates on the pragmatic aspect and is relieved of the need to fully understand what happens on the semantic level during the therapy process.

Interestingly, a shift in meaning can be observed with regard to various relevant terms. Chaos, healing and control are three examples of words which are now understood differently than was the case only a few years ago. Chaos, with its negative everyday life connotations, has become acceptable in scientific circles. The terms "active" and "passive" are increasingly melting into the word "interactive", as a consequence of the inclusion of circularity and feedback loops. Controlling and being controlled combine to become regulation and self-control. The classic separation between the physician as the "doer" and the patient who "endures" can also be broken down by way of the interactivity between physician and patient, and through the inclusion of bio-resonance therapy. In this sense the therapist then becomes a companion, who helps the patient help himself, and who facilitates the patient's self-healing processes. The analogy between bio-resonance therapy (Ludwig 1994) and Pyragas' approach using feedback (equation (4)) goes rather deep. An accessible (electric) signal is taken from the patient and reintroduced in a different form. In a sense this re-fed signal is similar ("principle of similarity"), but not identical, to that which was taken.

This principle conjures up all kinds of "feedback therapies", such as those using the patient's own blood or urine. Here, too, the question poses itself, how something which is present in the body anyway, and which is even regarded as a harmful excretory product, can have a positive effect. Simply by virtue of certain substances (such as blood or urine) having been outside the body for a certain interval, a small change already takes place. In addition, the substance is re-introduced into the body from the outside back into the system in a non-inherent way. Thus the immune defence could be stimulated in an unusual way.

Thus we finally come to homoeopathy, which certainly fits into this frame of thought. We are assuming that the (suitably altered) water structure of a fluid homoeopathic solution corresponds to the patient's (pathologically altered) water structure (Kratky, 1997). In this case, the homoeopath replaces the feedback loop by finding the principle of similarity. Again, the question poses itself how this, coming from the outside, can suddenly work. The same answers as proposed above are possible. Firstly, we are dealing with a principle of similarity, and thus the discovery of a hypothetical, "exactly" suited homoeopathic remedy might not even be optimal. Secondly, it does, however, come into the body from the outside. Indeed, Sukul (1997) observed that certain homoeopathic remedies worked in mice when taken via the tongue, but not when injected into the peritoneum. Thus another circle closes, and it finally seems possible to look at different complementary medical approaches in view of the fact that they share the aspect of feedback.

References

Huebinger B. (1993), Die Steuerung deterministisch - chaotischer Bewegungen. Harri Deutsch, Thun.

Kratky K.W., ed. (1991), Systemische Perspektiven. Carl Auer, Heidelberg.

Kratky K.W. (1992), Chaos and Disorder. In: Self-Organization and Clinical Psychology (eds. W. Tschacher, G. Schiepek and E.J. Brunner). Springer, Berlin, pp. 88-101.

Kratky K.W. (1995), Systems Thinking, Systems Control, and Systems Regulation. In: Advances in Systems Studies, Volume II (ed. G.E. Lasker). The International Institute of Advances Studies in Systems Research and Cybernetics, Windsor, Ontario, Canada, pp.11-15.

Kratky K.W. (1996), Analysis and Control of Chaotic Systems - Two Sides of the Same Coin? In: Proc. 14th International Congress on Cybernetics (ed. J. Ramaekers). Association Internationale de Cybernetique, Namur, Belgium, pp. 548-553.

Kratky K.W. (1997), Comparison of Different Dilutions of the Same Substance. In: High Dilution Effects on Cells and Integrated Systems (ed. C. Taddei-Ferretti), in print.

Ludwig W. (1994), SIT - System-Informations-Therapie. Schwingungsmedizin in Theorie und Praxis. Spitta, Balingen.

Mahnke R., J. Schmelzer and G. Röpke (1992), Nichtlineare Phänomene und Selbstorganisation. Teubner, Stuttgart.

Ott E., C. Grebogi and Y.A. Yorke (1990), Controlling Chaos. Phys. Rev. Lett. 64, pp. 1196-1199.

Pyragas K. (1992), Continuous Control of Chaos by Self-Controlling Feedback. Physics Letters A170, pp. 421-428.

Sukul N.C. (1997), Interaction of a High Dilution of Agaricus muscarius L with dopamine agonists and antagonists in modulating catalepsy of mice. In: High Dilution Effects on Cells and Integrated Systems (ed. C. Taddei-Ferretti), in print.

BIOLOGICAL EFFECTS OF ELECTROMAGNETIC FIELDS

Paolo Bellavite and Andrea Signorini

Molecular biology today appears to be the mainstream interpretative basis for all cellular and pathophysiological phenomena, even going so far as to embrace neuronal and psychic events. The explanation of disease processes, whether genetic or acquired, is sought and, where possible, located in mechanisms consisting in quantitative and/or qualitative modifications of particular molecules making up part of the various anatomical or physiological systems. In relation with the molecular paradigm of disease, also the medical treatment is today founded mostly on the conventional pharmacological approach. However, the need of further research in different fields is suggested by the fact that biological phenomena are characterized by high levels of organization, in which inter-molecular and inter-cellular communications of biophysical nature appear to be involved according to many lines of evidence (Bellavite and Signorini, 1995). As suggested by Kroy (1989), in the living creatures there is an ancestral cybernetic order that is not based on the nervous system or on the humoral system (blood, hormones). This ancestral system is thought to be of an electromagnetic nature, because electromagnetic radiation is the most basic form of information carrier present in nature. Electromagnetic signals have constituted (and still constitute) both the language of communication between atoms and molecules and the means whereby primordial organisms received a series of items of information on the environment (sunlight, other cosmic waves). Living organisms have learnt to use electromagnetism as an information signalling system and even as a means of communication between cells and tissues. According to the studies by Popp and coworkers (Popp 1985; Popp *et al.* 1989; Ho and Popp 1994), many biological systems are capable of producing, receiving and even of "storing" electromagnetic waves such as light.

Moreover, considering most of major health problems such as neoplastic, degenerative, auto-immune, endocrine-metabolic, and neuropsychiatric diseases, it should be noted that they can not fully explained using a strictly reductionistic molecular paradigm, because they are not due to single modifications of specific genes/molecules, but often to subtle and minor changes of genetic predisposition to disease, which in turn react with a multiplicity of environmental factors to determine the

J. Schulte and P.C. Endler (eds.),
Fundamental Research in Ultra High Dilutionand Homoeopathy, 127-142.

final risk to develop disease and symptoms. Therefore, the understanding of the organization of the systems regulating these interactions is needed in order to properly deal with these diseases. The pursuit of a unitary approach, capable of assuring a multidisciplinary synthesis and of defining the nature of disease processes at higher levels of organization/dis-organization, is proving increasingly difficult using the molecular paradigm only. In practice, though not in principle, the ongoing accumulation of notions and data regarding single genes and molecules proves insufficient to the purpose of understanding complex vital phenomena and thus of the phenomena relating to health and disease.

In this context, the emerging bioelectromagnetic paradigm will play an important role because it re-evaluates important forms of long-range communication not only at inter-molecular but also at supra-molecular levels of complexity of biological systems. Moreover, a biophysical paradigm is necessary in order to develop working models and hypotheses explaining the possible effect of highly diluted remedies, that are often used by homoeopathic doctors at dilutions far beyond the Avogadro number. Only very recently first tenable steps have been taken to explain how a transfer of information from such remedies to the body can take place (this volume will hopefully stimulate further development). It has been recognized that in solving the problem explanations on serious physics grounds are required (not merely chemistry). It is very likely that any such explanation must necessarily take account of sensitivity of living systems to low energy electromagnetic fields.

The study of the effects of electromagnetic fields in the body has come to take on increasing importance and scientific dignity in recent years, while at the same time that aura of mystery which has favored the exploitation of such phenomena by charlatans has steadily declined thanks to a well structured initiative of established researchers (see Linde this volume). The renewed interest in interactions between electromagnetic fields and living systems consist in various factors, including the following: a) evidence has been building up regarding the efficacy of extremely-low-frequency (ELF) pulsed magnetic fields in therapy, and most notably in orthopaedics; b) from the standpoint of public health, there is a heightened awareness of the risks associated with technological development and thus also with exposure to electromagnetic fields generated, for instance, by high-voltage electrical power lines, and other sources; c) the topic is being tackled increasingly in experimental terms with studies on cell and molecular models, with the result that a number of possible explanations of the biological effects of low-energy magnetic fields are beginning to emerge.

These matters will be briefly illustrated here below, as a contribution to a better understanding of the emerging biophysical paradigm in medicine and thus of the possible relationship between electromagnetic phenomena and homoeopathy. We feel we should stress that our discussion of these issues lays no claim to being in any way systematic, but rather constitutes an attempt to compare in outline and to put many different problems and phenomena into perspective which are sill largely unclear from the scientific point of view.

Electromagnetic fields

For the purposes of making it easier to understand the basic concepts used in bioelectromagnetism and the experimental evidence reported here below, we will briefly explain the terminology and the measurement units used. A diagram illustrating the various types of electromagnetic waves, together with their wavelengths and frequencies is given in Figure 1.

The frequency of an electromagnetic field is the number of cycles per second of the electromagnetic wave, or the number of pulsations of the field itself per second and is measured in Hertz (Hz). The wavelength (λ) is the distance between two wave peaks and is measured in meters (or in multiples of submultiples of a meter). Obviously, the higher the frequency, the lower will be the wavelength.

Electromagnetic waves are used, as is known in the case of telecommunications, as information vectors. For this purpose a carrier wave is used with a frequency selected in a very broad range according to the transmission and reception systems. This carrier wave is specifically *modulated* in relation to the information to be conveyed, i.e. its length and height are subtly altered and can be slightly increased or reduced to a variable extent over time (frequency and amplitude modulation, respectively). In this way, a piece of equipment *tuned* to the carrier wave can perceive the modulation and, after decoding it, the information contained in it.

The *intensity* of the electrical field is provided by the electric potential over a given distance and is expressed in volts/meter (V/m) or millivolts/centimeter (mV/cm). When a biological system is exposed to an electrical field, the mobile charges shift in the

Figure 1

	frequency	wavelength	time scale	biological phenomena	physical	phenomena
	10^{20}				γ-ray	
	10^{19}				hard/soft	
	10^{18}	300 pm			X-ray	ionizing
	10^{17}				UV	
	10^{16}					
PHz	10^{15}	300 nm		biophotons	visible light	
	10^{14}				IR	
	10^{13}					
THz	10^{12}	300 µm				
	10^{11}				microwave	non-ionizing
	10^{10}				radar	
GHz	10^{9}	300 mm			radar	
	10^{8}			electrochemical cell metabolism	video, FM	
	10^{7}			electromag. cell-to-cell communication	VHF,short wave	
MHz	10^{6}	300 m			AM	
	10^{5}				long wave	
	10^{4}				induction heating	
kHz	10^{3}	300 km	ms	neural activation		
	10^{2}				power grid, ELF	
	10^{1}			EEG (βαδ)		
Hz	10^{0}	3×10^{8}m	s	heartbeat		
	10^{-1}			respiration		
	10^{-2}		min			
mHz	10^{-3}	3×10^{11}m				
	10^{-4}		h			
	10^{-5}		d	biol. rest/activity cycle		
µHz	10^{-6}	3×10^{14}m	month	mentrual cycle		
	10^{-7}		yr			
	10^{-8}			human popul. cycle		
pHz	10^{-9}	3×10^{17}m	human genration			
	10^{-10}					

Courtesy J Schulte, UTS

direction induced by the field itself, thus forming a *current*, which is measured in amps (A) or in submultiples of an amp. With reference to a certain area of tissue or organ traversed by electrical charges, there will be a certain density (j) of the current itself, which is measured in amps/square meter A/m^2.

The electric field and the magnetic field are closely related according to Faraday's law of induction. When a pulsed magnetic field is applied to an electrically conducting material (such as living matter), an electric field is introduced perpendicular to the direction (vector) of the magnetic field. This electric field obviously depends on the surface of the area concerned and is proportional in intensity to the frequency of the magnetic field and its intensity.

The intensity of the magnetic field is measured in Gauss (G) or, to use the more modern SI unit, in tesla (T) or submultiples of a tesla ($1T = 10^4G$). To have two terms of comparison, the intensity of the earth's magnetic field is of the order of 0.02 to 0.07 mT (0.2 to 0.7 G), whereas that used in diagnostics by magnetic resonance is of the order of 0.1 to 10 mT (1 to 100 G) (Walleczek 1992).

Effects on the organism

We intend here to examine low-energy, low-frequency radiation, which acts with very different mechanisms compared to ionizing radiation. The latter causes biological effects through ionization (detachment of electrons from the atomic orbits) of the molecules and thus gross alterations such as damage to chromosomes, peroxidation of lipid membranes, and so on. In contrast, the energy of radiation with frequencies from 0 to a few hundred GHz is too low to cause physicochemical changes of this type and at most is able to yield thermal effects (heating, used, amongst other things, in the functioning of microwave ovens).

The effects of non-ionizing electromagnetic fields on the human body may be both of a pathological type and useful for therapeutic purposes. As regards the damaging effects most commonly studied, we have to refer essentially to studies which appear to demonstrate an increase in tumours in exposed subjects (Pool 1990). The topic is much debated and the epidemiological data have been confirmed only with regard to a number of childhood tumours (leukemia). As regards the uses for therapeutic purposes, the techniques most extensively employed are electromagnetic stimulation of osteogenesis, in cases of pseudoarthrosis and retardation of fracture consolidation (Chiabrera *et al.*, 1984). This is not the place for a detailed review of the pathological or therapeutic effects of electromagnetic fields, this today being an area of major development, and so we will confine ourselves to outlining the basic molecular and cellular aspects.

There are many natural sources of weak electromagnetic fields: sources outside the body include, for instance, the earth's magnetic field (which is exploited by a number of birds, fish and dolphins for direction-finding), signals from the earth's ionosphere (Schumann waves at 7.8 Hz that are related to the accuracy of biological rhythms), radiation from the stars which emit radiofrequencies, the sun itself (particularly in certain phases of its activity) (Konig 1989), the waves irradiated by telecommunications and radar systems, and electrical power lines. The sources inside the body are multiple and range from the electrical activity of the brain (e.g. hippocampus at 7.8 Hz), the nerves and muscles to the electric fields generated by a number of fish and other marine organisms (used for the purposes of recognition in the dark and for defence), to the generation of light by cells such as leucocytes (chemiluminescence).

Practically all organisms emit light at a rate from few photons per cell per day to several hundred photons per organism per second. This emission of biophotons, as they are called, is distinct from chemiluminescence of leucocytes and of bioluminescence of fireflies, that is associated with specific organelles. Biophoton emission occurs at very low intensity but is universal to living organisms, where is thought to represent a long-range form of communication, capable of generating synchronous and coherent phenomena (for a review, see Ho and Popp, 1994).

The electrocardiogram and electroencephalogram are no more than two methods of measuring the endogenous electrical activity of the heart and the nerve centres. Electrical activity is also generated in bone when it is deformed; such activity can be defined as piezoelectric and appears to be important for directing the growth of bone trabeculae along lines of force. In actual fact, one of the first clinical uses of weak magnetic fields was precisely the induction of bone repair (Bassett *et al.* 1974).

Animal organisms have developed very marked sensitivity to electromagnetic waves. Without going beyond the most obvious field, we need only mention the sensitivity of the eye to light, which makes it capable of perceiving only a few photons.

The experiments by Smith and Monro (Smith *et al.* 1985; Monro 1987; Smith 1988; Smith 1989, 1994) illustrate the concept of "sensitivity" to minimal perturbations of electromagnetic fields. Smith et al. reported a series of experiments performed in collaboration with allergologists from Hospitals in London and Dallas, in which they succeeded in inducing allergic manifestations in patients with immediate hypersensitivity to many substances, simply by bringing them into close contact with sources of electromagnetic radiation. The allergic manifestations could set in rapidly at particular frequency bands ranging, according to the individual patients, from only a few mHz to a large number of MHz. It was not, then, so much the intensity of output of the oscillator (a few V/m) that was important as the *frequency* and *coherence*.

It is not only curious that these investigators demonstrated the ability to trigger allergic attacks with electromagnetic waves, but also that the patients sensitive to this type of stimulation themselves produced electromagnetic signals during the allergy attacks, though the latter were provoked chemically. For details, see the contribution Endler et al. in this volume.

It has been demonstrated that a number of species of fish are capable of perceiving and responding to electric fields with intensities as low as 1 μV/m (Bullock 1977), which corresponds to the most marked sensitivities found in allergic subjects. Again according to Smith, such sensitivities may enable the fish to locate food at great distances: it has been seen, in fact, that living cells, such as, for instance, yeasts, emit electromagnetic waves in radiofrequencies at levels of approximately 0.1 V/m (Smith 1988; Pollock and Pohl 1988).

In the course of the allergometric tests in the sample of hyperreactive patients, the researchers realized that allergic reactions triggered by contact with chemical agents could be neutralized by treating the patients with particular frequencies. If the same frequencies were used to treat pure water, the latter acquired the neutralizing therapeutic properties. If, on the other hand, the water was exposed to frequencies capable of triggering the attack, it acquired the properties of an allergen (for respective experiments with amphibia, see Pongratz and Endler this volume)

The treatment of the water was done by inserting glass test tubes containing the water in solenoids or thoroids powered by an oscillator. The changes induced in the water, capable of triggering allergic attacks in hypersensitive patients, persisted for 1-2 months. At this point it is interesting to note that the stability of the homoeopathic remedy in aqueous solution is traditionally short-lived, of the order of months, whereas since Hahnemann in homoeopathy water-alcohol solutions are used precisely because they were much more stable and long-lasting (years).

Quite apart from the fact that only a minority of allergic patients exhibited this extreme sensitivity and were suitable for the execution of such tests, the demonstration of the ability of water to incorporate electromagnetic information and transmit it to individuals reactive to it remains, if independently repeatable, of great interest and significance. The respective multicenter experiments of Endler *et al.* are described in detail in this volume. In this study, amphibian larvae were exposed to thyroxine dilutions prepared according to a homoeopathic protocol, whereby the liquid (as well as water control) were sealed in glass vials that were brought into the basin water containing the amphibian larvae. Significant effects on the metamorphosis rate, similar to those of the liquid added directly dropwise to the water, were observed. This seems to be a valid argument in favor of the effective existence of metamolecular information and its presence in aqueous solutions

Cellular effects

It is well known that electromagnetic radiation can cause substantial changes at cellular level, but the bulk of attention, up until not very long ago, was devoted to the potentially toxic effects of medium-to-high energy radiation, such as ultraviolet light, gamma and X-ray. As mentioned above, investigations into the mechanisms of the biological effects of non-ionizing radiation have begun recently (Figure 1).

Electromagnetic waves, even if of low energy and broad wavelength, are known to generate heat, when absorbed by biological matter. The question whether millimetre waves cause effects independent of absorption of heat, i.e. so-called nonthermal effects, has been the subject of lengthy scientific debate. The controversy regarding the existence of cell responses to low-energy waves is due both to the fact that the reproducibility of many experiments has proven difficult, and to theoretical objections that the energy of such weak fields would be less than the energy of the background noise due to the temperature at which the cells are studied (*thermal noise*). If we are to expect an effect of an electromagnetic field applied from the outside, this field will have to cause significantly greater changes than would in any event occur casually in biological systems even in the resting state (e.g. the continual opening and closing of ion channels, oscillations in membrane potentials and in many metabolic activities, etc., all these being processes which are in any event active at a certain temperature). Today, however, the existence of nonthermal effects of weak electromagnetic fields has been demonstrated in many experimental systems and may now be regarded as generally accepted (Kremer *et al.* 1988; Aldrich and Easterly 1987; Magnavita 1989; Tsong 1989).

A major contribution to this issue can be found in a critical study published in *Science* (Weaver and Astumian 1990). These authors propose physical models according to which the cells are considered as detectors of very weak periodic magnetic fields and where the relationships between the size of the cell and the changes in membrane potential due both to temperature-induced fluctuations and to the application of electromagnetic fields are established. In the simplest version of the model, the calculation estimates at around 10^{-3} V/cm the intensity of the minimum field to which the membrane macromolecules could be sensitive. However, if the model parameters considered take into account the so-called frequency "windows", i.e. the possibility that certain responses occur only within a restricted frequency band, then the theoretical intensity necessarily proves to be several orders of magnitude lower (μV/cm), thus closely approaching the data from various experiments in cells and animals.

The growth of the nerve processes is guided by weak electric currents (Alberts *et al.*, 1989). When a nerve process lengthens in culture or even in connective tissue, at its apex a structure called a growth cone is formed, which appears as the expansion center of many long filaments (filopods) which look like continually slow-moving finger-like processes, making ameboid movements: some retract, and others stretch out, as if exploring the terrain. Within the filopods many actin filaments are to be found. The net vectorial shift of the growth cone in one direction is followed by a lengthening of the nerve fibre (at an estimated rate of 1 mm per day). The direction of the movement depends on various local factors, such as, for instance, the orientation of the fibre of the connective tissue matrix, along which the growth preferentially occurs, and even the existence of specific membrane recognition systems between adjacent cells. The cells, however, are also powerfully influenced by electromagnetic fields: the growth cones of neurons in culture are oriented and direct themselves towards a negative electrode in the presence of low-intensity fields (70 mV/cm).

The cells have an ability to receive and integrate light signals, perceiving both their frequency and direction. This has been demonstrated by means of special phase-contrast microscopy equipment with infra-red light (Albrecht-Buehler 1991). 3T3 fibroblasts in culture extend the filopods preferentially towards light sources, the most effective being the intermittent ones in the 800-900 nm range with 30-60 impulses per minute. According to the author of these experiments, the cell receptor for the radiation is the centrosome.

There is also evidence that cell proliferative activity is influenced by electromagnetic fields, albeit of very low intensity (0.2-20 mT, 0.02-1.0 mV/cm) (Luben *et al.* 1982; Conti *et al.* 1983; Cadossi *et al.* 1992; Walleczek 1992).

It is important to note that on the basis of the literature data available to date it is impossible to draw any definite conclusions as to the positive or negative, stimulatory or inhibitory effects of weak electromagnetic fields on cellular or molecular systems and above all as to doses and application modalities (Walleczek 1992). In fact, the bioactive electromagnetic signals used vary very considerably in terms of intensity, frequency, duration, and waveform (sinusoidal, square, sawtooth, etc.). Moreover, the effect may also depend on the biological status of the cells exposed (Cossarizza *et al.* 1989; Walleczek and Liburdy 1990), indicating that mechanisms of very complex interaction between various different factors are involved.

Many enzymes and receptors appear to be sensitive to stimulations of a physical as well as a chemical type (Adey 1988; Tsong 1989; Popp *et al.* 1989). The cell membrane, by virtue of its bioelectrical properties, is the site where influences of this type are most likely to be exerted (Kell 1988), though other possible candidates are the large macromolecules organized in repetitive units, such as the nucleic acids (Popp 1985), or the proteins of the cytoskeleton, particularly the microtubules (Hameroff 1988).

The biological basis of the effect of magnetic fields on cells is highly complex and cannot be analyzed exhaustively here. The cell constitutes a typical electrochemical system, with a transmembrane potential difference (negative outside compared to inside) and a very large number of proteins endowed with electric charges of varying sign. According to the fluid mosaic model of the membrane (a model which is still valid, at least in general terms) in an ideal cell at rest, the proteins are distributed evenly over the membrane, but, in the presence of an electric field crossing the membrane, they undergo electrophoretic attraction or repulsion, tending to shift towards the poles which the cell presents in the direction of the electric field. A current of electrons or ions invading a cell flows around it, causing a movement of (electrically charged) proteins in the opposite direction.

The rearrangement of the position of the proteins on the surface of the membrane is not devoid of consequences, in that it favors contact between neighboring proteins and slows down contact between distant proteins (Chiabrera *et al.* 1984). Since the functioning of receptors and membrane transduction systems depends on aggregations of proteins or at least on contacts between proteins, the consequences of the electric field for cell activation are easily imaginable. The aggregation phenomenon normally occurs in the case of a chemical signal, because the signal molecule may serve as a bridge between two or more receptors, which are mobile in the plane of the membrane. Of course, this model is a very substantial simplification of what happens in reality, where the concentrations of calcium, magnesium, sodium, potassium and hydrogen ions come into play, as well as the possible direct effect of the magnetic field on the macromolecules of enzymes, receptors or the cytoskeleton.

Across the double lipid layer of the biological membranes, measuring approximately 40 Å in thickness, an electrical gradient of a few tens or hundreds of mV is established, which means something like 10^5 V/cm. Theoretically this gradient should constitute an effective electrical barrier against minimal perturbations such as those created by low-frequency electromagnetic fields present in the extracellular membrane. In other words, the natural electrical activity of the membrane would constitute a kind of "background noise" which would prevent the possibility of perceiving minimal variations in potential. Recent research, however, has shown that electromagnetic fields various orders of magnitude weaker than the transmembrane potential gradient are capable of modulating the actions of hormones, antibodies and neurotransmitters at receptor and transduction system level. This suggests that highly cooperative processes are set up, i.e. that repeated minimal variations cooperate to cause major effects (for a more detailed treatment of cooperative processes see Schulte this volume).

The sensitivities observed in these biological processes of electromagnetic modulation are of the order of 10^{-7} V/cm in the E.L.F. (extreme low frequency) range.

Note, for example, that electric phenomena responsible for the EEG create gradients of 10^{-1} to 10^{-2} volts/cm (Adey 1988). Moreover, many of these interactions depend on the frequency, i.e. they occur only in certain *windows of frequency*, which would suggest the existence of *nonlinear* regulation systems far from equilibrium (Adey 1988; Weaver and Astumian 1990; Yost and Liburdy 1992). Similar sensitivities have been detected in a broad spectrum of tissues and cells, indicating that we are faced with a general biological property characteristic of cells. Furthermore, also *windows of the intensity* of the field are suggested (Pongratz *et al.* this volume).

Molecular effects

It is known that many molecular elements with receptor, structural and enzymatic functions are sensitive to changes in weak electromagnetic fields: photoreceptors (Alberts *et al.* 1989), chlorophyll (Alberts *et al.* 1989), receptors with 7 trans-membrane domains (Bistolfi 1989), G-proteins (Adey, 1988), cAMP-dependent protein kinase (Byus *et al.* 1984), protein kinase C (Adey 1988), lysozyme (Shaya and Smith 1977), receptors (aggregation) (Chiabrera 1984), chromosomes (Kremer *et al.* 1988), protein and lipid biopolymers (Hasted 1988), Na^+/K^+ ATPase (Liu *et al.* 1990). Experimental data regarding these molecular systems will be detailed below.

Most protein molecules are capable of passing reversibly from one conformational state to another by virtue of various possible combinations of hydrogen bonds, disulfide bridges and hydrophobic forces. These passages occur by means of nonlinear changes, or hopping, to overcome the energy barriers between one state and another. The proteins are thus dynamic, vibrating structures whose components undergo continual oscillatory movements, which take place over a time scale ranging from femptoseconds (10^{-15} s) to several minutes. The most significant vibrations in biological systems are of the order of nanoseconds (Hameroff 1988). It is very important to stress the fact that, in biology, many proteins (and also other chemical species such as lipids) are assembled in multimeric or polymeric groups. In these structures, cooperative, or collective, interactions easily occur, with the result that the vibrations may propagate themselves in *coherent* ways and, as such, may take on a biological-informational significance (Frohlich 1988; Del Giudice *et al.* 1988b; Bistolfi 1989; Ludwig 1994; Del Giudice and Preparata, Schulte, this volume).

The transfer of both chemical and electromagnetic signals from the external surface of the cell across the membrane consists in the transmission of conformational variations and oscillatory motions of proteins which have transmembrane domains (segments of the molecule). It has been claimed that a key role in this transmission is played by portions of proteins that have helical or folded-sheet-shaped fibrous structures (Bistolfi 1989; Meissmer, cited from Ludwig 1994). Such structures are characterized by a substantial degree of order and by arrangement in repetitive sequences, as well as by the existence of hydrogen bonds between the amine residues of adjacent amino acids arranged longitudinally along the fibre. These protein structures are characteristic in their ability to *resound* according to nonlinear modes of vibration as a result of interaction with electromagnetic fields.

The prototype of this type of receptor is rhodopsin, the light receptor in the retina, which consists of 7 a-helixes arranged in orderly fashion transverse to the plane of the membrane on which it is situated. In this type of receptor-transducer, the excitation resulting from absorption of the photon is linked to the pumping of a proton and to the stabilization of a transmembrane potential.

It should be noted, however, that this structure with 7 a-helixes crossing the membrane is also found in an extensive family of glycoproteins involved in cell transmission systems coupled to G-proteins: the b-adrenergic receptors, the muscarinic receptors for acetylcholine, various receptors for neuropeptides, the receptors for chemotactic peptides in the white blood cells and even the mutual recognition systems in yeast cells involved in replicative fusion (Alberts *et al.* 1989). It is therefore likely that these characteristic structural features render the transmission systems they are present in susceptible to electromagnetic modulation.

Studies conducted on electromagnetic modulation of collagen production by osteoblasts are consistent with this view. It has been demonstrated, in fact, that parathyroid hormone in osteoblasts binds to external receptors and activates the enzyme adenylate cyclase via the mediation of a G-protein. An electromagnetic field with a 72 Hz frequency and an electrical gradient of 1.3 mV/cm induced 90% inhibition of adenylate cyclase activation without interfering either with the receptor binding or with the enzyme itself. As a result, the inhibitory effect was attributed to blockade of the G-protein (Adey 1988).

Cyclic AMP (cAMP) is an important element in controlling the function of many enzymes, in particular as an intracellular increase in cAMP constitutes an activatory message for the protein kinases (enzymes which phosphorylate proteins). In precise experimental conditions of frequency and duration of exposure, the cAMP-dependent protein kinase of human lymphocytes has been inhibited by electromagnetic waves (field of 450 MHz modulated in amplitude to 16 Hz). Type C protein kinase, the involvement of which in important cell processes as well as in carcinogenesis is beyond doubt, can also be modulated by electromagnetic waves (data from Byus, cited in Adey, 1988).

The catalytic activity of the enzyme lysozyme is sensitive to electromagnetic waves (radiofrequencies from 0.1 to 150 MHz) (Shaya and Smith 1977). In these experiments, solutions of lysozyme were exposed, in the presence of submaximal doses of the competitive inhibitor n-acetyl glucosamine (NAG), to various electromagnetic frequencies supplied by an oscillator by means of a coil wrapped around a polycarbonate container of the enzyme solution. The main effect observed was a modification of the inhibition produced by NAG. Interestingly, specific frequencies (e.g. 40 MHz) increased the effect of the inhibitor, and other frequencies (e.g. 100 MHz) decreased the effect, enhancing the activity to the level of the uninhibited lysozyme, while yet other frequencies (e.g. 150 MHz) had no effect. Inspection of the whole range of frequencies between 0.1 and 150 MHz showed alternating peaks of stimulation and inhibition of the enzyme activity, without any apparent regularity. Subsequent measurements between 30 MHz and 50 MHz showed further fine details in the effects produced. Therefore, the relationship between frequency and activity appears to show a chaotic trend and fractal behavior.

According to Tsong and coworkers (Tsong 1989; Liu *et al.* 1990), the conventionally known forms of intercellular communication, such as ligand-receptor interaction, are slow processes operating over short distances, but cells also need rapid forms of communication over long distances, with the result that it has been postulated that the various biochemical reactions, which are in any event necessary, are regulated by forces of a physical nature. Given that oscillating weak electromagnetic fields are capable of stimulating or suppressing many cell functions and that, from the thermodynamic point of view, it is argued that this is possible only if a mechanisms of amplification of the signal exist. Schulte (this volume) shows, however, that when considering coherent photon-phonon coupling an artificial amplification mechanism may not be required.

The experiments carried out be Tsong's team indicate that a weak electric field (20 V/cm) is capable of activating the function of Na^+/K^+-dependent ATPase only if specific frequencies are simultaneously used, corresponding to 1 kHz for the pumping of K^+ and 1 MHz for the pumping of Na^+. These results have led to the formulation of the concept of "electroconformational coupling". This model postulates that an enzymatic protein undergoes conformational changes as a result of a Coulomb interaction with an electric field (or with any other oscillating force field with which the protein can interact). When the frequency of the electric field corresponds to the characteristic kinetics of the conformational transformation reaction, a phenomenological oscillation is induced between different conformations of the enzyme. At the optimal field force, the conformations thus achieved are functional and the oscillations are utilized to perform the activity required, such as, for example, the pumping of Na^+ and K^+.

The organization of DNA in the chromosomes is affected by influences of an electromagnetic nature, as demonstrated in an extensive series of studies by Kremer and his coworkers (Kremer *et al.* 1988). These authors used the model provided by giant chromosomes of insects (larvae of *Acricotopus lucidus*), which are easily visible and can be studied under the microscope. It is well known that when information has to be transcribed from DNA to RNA, the chromosomes (compact rods containing thousands of genes packed and stabilized by isotonic proteins) have to partially decondense, showing puffs of genetic material issuing from the rod in the relevant segment. This phenomenon is strongly and significantly inhibited - in the sense that the puffs are much smaller - by irradiation of the chromosome with frequencies of around 40 to 80 GHz and outputs of only 6 mW/cm^2. The nonthermal nature of the phenomenon has been demonstrated by many control experiments. It is worth pointing out that even DNA and RNA are characterized by a macromolecular organization that is extremely rich of hydrogen bonds, connecting complementary nucleotide pairs, a structure that makes them good candidate for resonating events.

In this quite recent field of investigation, many points remain still to be clarified and any conclusion, particularly in the field of therapeutical applications, should be regarded as hypothetical and preliminary. However, present knowledge allows us to suggest that this kind of ordered communication network of biophysical nature, coupled with the well known high sensitivity of complex and chaotic systems to small perturbations, could be a physiological basis of the interaction between the body and information carried by low-doses or high-dilutions of pharmacologically active compounds and conceivably by low-energy electromagnetic fields.

References

Adey, W.R. (1988) Physiological signalling across cell membranes and cooperative influences of extremely low frequency electromagnetic fields. In: *Biological Coherence and Response to External Stimuli* (H. Frohlich, ed.) Springer-Verlag, Berlin, p. 148.

Alberts, B., Bray, D., Lewis, J., Raff, M., Roberts, K., Watson, J.D. (1989) *Molecular Biology of the Cell. 2nd edition.* Garland Publ., New York.

Albrecht-Buehler, G. (1990) In defense of "nonmolecular" cell biology. *Int. Rev. Cytol.* 120: 191.

Albrecht-Buehler, G. (1991) Surface extension of 3T3 cells towards distant infrared light sources. *J. Cell. Biol.* 114: 493.

Aldrich, T.E., Easterly, C.E. (1987) Electromagnetic fields and public health. *Environ. Health Perspect.* 75: 159.

Bassett, C.A.L., Pawluk, R.J., Pilla, A.A. (1974) Augmentation of bone repair by inductively coupled electromagnetic fields. *Science* 184: 575.

Bellavite, P., Signorini, A. (1995) *Homeopathy, A Frontier In Medical Science. Controlled Studies And Theoretical Foundations.* North Atlantic Books, Berkeley. Italian edition: IPSA Ed., Palermo

Bistolfi, F. (1989) *Radiazioni Non Ionizzanti, Ordine, Disordine e Biostrutture.* Edizioni Minerva Medica, Torino.

Byus, C.V., Lundak, R.L., Fletcher, R.M., Sadey, W.R. (1984) Alterations in protein kinase activity following exposure of cultured lymphocytes to modulated microwave fields. *Bioelectromagnetics* 5: 34.

Bullock, T.H. (1977) Electromagnetic sensing in fish. *Neurosci. Res. Program. Bull.* 15: 17.

Cadossi, R., Bersani, F., Cossarizza, A., Zucchini, P., Emilia, G., Torelli, G., Franceschi, C. (1992) Lymphocytes and low-frequency electromagnetic fields. *FASEB J.* 6: 2667.

Chiabrera, A., Grattarola, M., Parodi, G., Marcer, M. (1984) Interazione tra campo elettromagnetico e cellule. Le Scienze (Italian edition of Sci. Am.) 192: 78.

Conti, P., Gigante, G., Cifone, M.G., Alesse, E., Ianni, G.F., Reale M., Angeletti, P.U. (1983) Reduced mitogenic stimulation of human lymphocytes by extremely low frequency electromagnetic fields. *FEBS Lett.* 162: 156.

Cossarizza, A., Monti, D., Bersani, F., Cantini, M., Cadossi, R., Sacchi, A., Franceschi, C. (1989) Extremely low frequency pulsed electromagnetic fields increase cell proliferation in lymphocytes from young and aged subjects. *Biochem. Biophys. Res. Commun.* 160: 692.

Del Giudice, E., Doglia, S., Milani, M., Vitiello, G. (1988b) Structures, correlations and electromagnetic interactions in living matter: Theory and applications. In: *Biological Coherence and Response to External Stimuli* (H. Frohlich, ed.). Springer-Verlag, Berlin, p. 49.

Frohlich, H. (ed.) (1988) *Biological Coherence and Response to External Stimuli*. Springer-Verlag, Berlin.

Hameroff, S.R. (1988) Coherence in the cytoskeleton: Implications for biological information processing. In: *Biological Coherence and Response to External Stimuli* (H. Frohlich, ed.). Springer-Verlag, Berlin, p. 242.

Hasted, J.B. (1988) Metastable states of biopolymers. In: *Biological Coherence and Response to External Stimuli* (H. Frohlich, ed.). Springer-Verlag, Berlin, p. 102.

Ho, M.W., Popp, F.A. (1994) Biological organization, coherence, and light emission. In: *Thinking About Biology* (W. Stein and F.J. Varela, eds.). SFI Studies in the ciences of Complexity, Lect. Note vol. III, Addison-Weisley Pub. Comp., Reading, MA, p. 183.

Kell, D.B. (1988) Coherent properties of energy-coupling membrane systems. In: *Biological Coherence and Response to External Stimuli* (E. Frohlich, ed.). Springer Verlag, Berlin, p. 233.

Konig, H.L. (1989) Bioinformation. Electrophysical aspects. In: *Electromagnetic Bio-Information* (Popp, F.A. *et al.*, eds.) Urban & Schwarzenberg, Munchen, p. 42.

Kremer, F., Santo, L., Poglitsch, A., Koschnitzke, C., Behrens, H., Genzel, L. (1988) The influence of low-intensity millimeter waves on biological systems. In: *Biological Coherence and Response to External Stimuli* (H. Frohlich, ed.). Springer-Verlag, Berlin, p. 86.

Kroy, W. (1989). The use of optical radiation for stimulation therapy. In: *Electromagnetic Bio-Information* (Popp, F.A. *et al.*, eds.). Urban & Schwarzenberg, Munchen, p. 200.

Liu, D.S., Astumian, R.D., Tsong, T.Y. (1990) Activation of Na^+ and K^+ pumping modes of (Na,K)-ATPase by an oscillating electric field. *J. Biol. Chem.* 265: 7260.

Luben, R.A., Cain, C.D., Chen, M.Y., Rosen, D.M., Adey, W.R. (1982) Effects of electromagnetic stimuli on bone and bone cells in vitro: inhibition of responses to parathyroid hormone by low-energy, low-frequency fields. *Proc. Nal. Acad. Sci. USA* 79: 4180.

Magnavita, N. (1989) Radiazioni elettromagnetiche e rischi per la salute. *The Pratictioner (Italian edition)* 124: 30.

Pollock, J.K., Pohl, D.G. (1988) Emission of radiation by active cells. In: *Biological Coherence and Response to External Stimuli* (H. Frohlich, ed.). Springer-Verlag, Berlin, p. 139.

Pool, R. (1990) Is there an EMF-cancer connection? *Science* 249: 1096.

Popp, F.A. (1985) *Nuovi Orizzonti in Medicina. La teoria dei Biofotoni.* IPSA Editore, Palermo.

Popp, F.A., Warnke, U., Konig, H.L., Peschla, W. (1989) (editors) *Electromagnetic Bio-information.* Urban and Schwarzenberg, Munchen.

Shaya, S.Y., Smith, C.W. (1977) The effects of magnetic and radiofrequency fields on the activity of lysozyme. *Collective Phenomena* 2: 215.

Smith, C.W., Choy, R., Monro, J.A. (1985) Water - friend or foe? *Lab. Pract.* 34: 29.

Smith, C.W. (1988) Electromagnetic effects in humans. In: *Biological Coherence and Response to External Stimuli* (H. Frohlich, ed.). Springer-Verlag, Berlin, p. 205.

Smith, C.W. (1989) Coherent electromagnetic fields and bio-communication. In: *Electromagnetic Bio-Information* (Popp, F.A. *et al.*, eds.). Urban and Swarzenberg, Munchen, p. 1.

Smith, C.W. (1990) Homeopathy, structure and coherence. In: *Homeopathy in Focus.* VGM Verlag fur Ganzheitsmedizin, Essen, p. 96.

Smith, C.W. (1994) Electromagnetic and magnetic vector potential bio-information and water. In: *Ultra High Dilution* (P.C. Endler and J. Schulte, eds.). Kluwer Acad. Publ., Dordrecht, p. 187.

Tsong, T.Y. (1989) Deciphering the language of cells. *Trends Biochem. Sci.* 14: 89.

Walleczek, J., Liburdy, R.P. (1990) Nonthermal 60-Hz sinusoidal magnetic-field exposure enhances $45Ca^{2+}$ uptake in rat thymocytes: dependence on mitogen activation. *FEBS Lett.* 271: 157.

Walleczek, J. (1992) Electromagnetic effects on cells of the immune system: the role of calcium signaling. *FASEB J.* 6: 3177.

Weaver, J.C., Astumian, R.D. (1990) The response of living cells to very weak electric fields: the thermal noise limit. *Science* 247: 459.

Yost, M.G., Liburdy, R.P. (1992) Time-varying and static magnetic fields act in combination to alter calcium signal transduction in the lymphocyte. *FEBS Lett.* 296: 117.

HIGHLY DILUTED AGITATED SILVER NITRATE AND WHEAT SEEDLING DEVELOPMENT

EFFECT KINETICS OF A PROCESS OF SUCCESSIVE AGITATION PHASES

Waltraud Pongratz, Andrea Nograsek, Christian Endler

Silver nitrate is known to be generally biostatic. However, it is also discussed as a biologically vital trace substance. This substance has been prepared in a standardized process of stepwise dilution and agitation. Its influence on the development of wheat seedlings has been studied. When seedlings are exposed to silver nitrate at high molecular concentration, growth is normally *inhibited.* Interestingly, growth was *stimulated* by a dilution log 24 of this probe. In our experiments, addition of one more standardized step of dilution and agitation, or agitation alone, to the dilution log 24, significantly diminished the effect of the test dilution, whereas addition of two steps again enhanced the stimulatory effect ("double switch" kinetics). This was found by different independent researchers. Furthermore, optimized conditions for performance of the experiment are discussed.

The effect of a substance which, in common biology and medicine, is known to be biostatic, but which is also discussed as a necessary trace substance, namely silver nitrate (Strasburger 1990, Haidvogel 1997), on the development of wheat seedlings was investigated. In a previous study, seedlings were exposed to silver nitrate at high molecular concentration (10 mg per ml water, dilution log 2); germination and growth were drastically *inhibited* (unpublished material). However, at lower concentrations (e.g. in a dilution log 7) silver nitrate is able to exert a stimulating effect on botanical systems [1]. In two multicenter studies, seedlings were exposed either to an aqueous test dilution log 24 of this molecular solution, specially prepared by stepwise dilution (23 steps 1 : 10) and vigorous agitation (24 agitation phases), called 'silver nitrate D24' or to analogously prepared solvent water (or to unprepared solvent water). In these studies, growth was *stimulated* by the test dilution with statistical significance. This was shown by four independent researchers including researchers from the Universities of Graz and of Vienna (Pongratz et al. 1991, 1994).

J. Schulte and P.C. Endler (eds.),
Fundamental Research in Ultra High Dilutionand Homoeopathy, 143-154.

The study presented in this paper is based on findings reported in the literature, namely that adding one more step of dilution and agitation to D24 (= D25) significantly *diminishes* the effect of the test dilution, whereas adding two more steps (= D26) again *enhances* the stimulating effect (Kolisko 1926). A 'double switch kinetics' - pattern was postulated (D24: significant stimulation with regard to reference - D25: less or no stimulation - D26: again significant stimulation). In this experiment, obviously, both water submitted to 24 agitation phases as well as unprepared water can be used as reference (unpublished data).

In the study presented here, the effect pattern of the three dilutions silver nitrate D24, D25 and D26 was observed in experiments at different times of observation. The aim of this was to learn about the conditions of the 'double switch effect' (see above) with regard to the organisms involved. Furthermore, starting from a dilution silver nitrate D24, the process of stepwise dilution and vigorous agitation was replaced by a process of stepwise agitation without dilution. The aim of this was to learn about the conditions of the 'double switch effect' with regard to the dilution involved.

Methods

Plants: Unbroken, unsorted grains of winter wheat (grown without an application of herbicides or pesticides) (Triticum aestivum, sorte: Mephisto) were used.

Observed development: It is reported that biological systems in circumscript developmental transitions show maximal sensitivity (Smith 1990. In our study, the initial development of stalks was observed.

Preparation of test solutions: The grains were observed under the influence of specially prepared aqueous solutions of silver nitrate (Merck). The solution that was used as the stock solution was also used in the previous experiments concerning the influence of the molecular substance on wheat seedlings (see introduction). It contained 10 mg silver nitrate per 1 ml distilled (not deionized) water, and it was diluted in distilled water in steps of 1 : 10 (parts of volume). The stock solution and each of the different successive dilutions 1 : 10 were vigorously agitated: at every step, a disposable sterile glass bottle with a volume of 20 ml, filled with 9 ml of distilled water was used. Each time 1 ml of the silver nitrate dilution from the previous step was added to this bottle, which then was pushed down 50 times at intervals of 1.2 seconds (= 1 minute) (agitation by hand, pushing down against the left palm) to create mechanical shocks. For the last steps of dilution, larger bottles were used, but the ratio [volume : content] was as above. After each dilution plus agitation - period, the solution was given a rest of 1.25 minutes. One test dilution was prepared with 24 steps of agitation (agitation of the stock solution plus agitation of 23 further dilutions 1:10) (called 'D24'), one with

25 steps ('D25') and one with 26 steps ('D26'). Annotation: due to a redactional error, the respective protocol in reference (Pongratz et al., 1994), p. 24, line 38 is incorrect. It should read "aqueous solution $1{:}10^{24}$ part of the weight of the stock solution of silver nitrate" as above.

Several analogously prepared sets of ranges of the dilutions were used.

In a further part of the study, starting from a dilution silver nitrate D24, the process of stepwise dilution and agitation was replaced by a process of stepwise vigorous agitation (without dilution). The agitation phases were always separated by resting phases. Three test substances were called silver nitrate D24/1 (which is similar to D24), D24/2 and D24/3.

Untreated solvent (water) was used as reference. All sets were applied blindly, i.e. the substances were all coded.

Laboratories involved: Experiments were performed in the laboratories of W. Pongratz, C. Endler and A. Nograsek. For further details, see the Results Section.

Circadian specificity: With regard to older literature, it was assumed that the experiments should be started at noon. Thus, most of the experiments were started between 8 a.m. and 1 p.m.

Exposition to probes and Data base: The grains with the germination furrow downwards were put on floss paper in sterile glass dishes (diameter 11cm). They were arranged in the form of a circle with all germination pools towards its center. This was done in order to make observation of the stalks easier. However, previous experiments have shown that also a random distribution of the grains can be used. Each dish contained 20 ml of the respective probe. It was covered by a glass beaker with a volume of 1000 ml. Dish and beaker were wrapped in stannium foil and kept at 20oC in a temperature stabilized chamber at. In one part of the study, the stalk lengths were measured after 7 days, in another part after 5 days. In the experiments, three sets of dishes were always used for the treatment with the silver nitrate dilutions plus one set for the water control. 20 - 30 seedlings were put into one dish; the respective number of grains per dish was equal for all groups (for total numbers, see below).

Evaluation of the data: The arithmetic mean was used to describe the stalk length, and the lengths were compared by one way analysis of variance (Fisher-test). 1 S.D. of the mean was calculated. The groups of seedlings treated with silver nitrate D24, D25 and D26 were compared to the water control. In addition to this, with regard to the hypothesis of a double switch kinetics, the groups treated with silver nitrate D24 and D26, respectively, were statistically compared to the D25 - group. Here, only results at the 1%-level were considered for interpretation. The D24/1, D24/2 and D24/3 - groups and water control were compared analogously.

Results

Experiments Set I, II and III on step by step diluted agitated substances

SET I: Observation after 7 days (W. Pongratz)

Experiments with 4 groups of 2140 grains each, treated with silver nitrate D24, D25, D26 and unprepared solvent, respectively, were performed. When the data of all experiments were pooled, the comparison of the mean stalk length showed a characteristic effect pattern (see introduction) of the three dilutions. The values for the silver nitrate - groups were generally above the control - value (see Figure 1). When the test substances were mutually compared, both the groups treated with the dilution D24 as well as with D26 showed more growth than the group treated with the dilution D25 (+ 4.3% and + 4.3%, respectively). This difference is statistically significant ($P < 0.01$ and < 0.01; for a comprehensive survey over P - values, see Tab. 5).

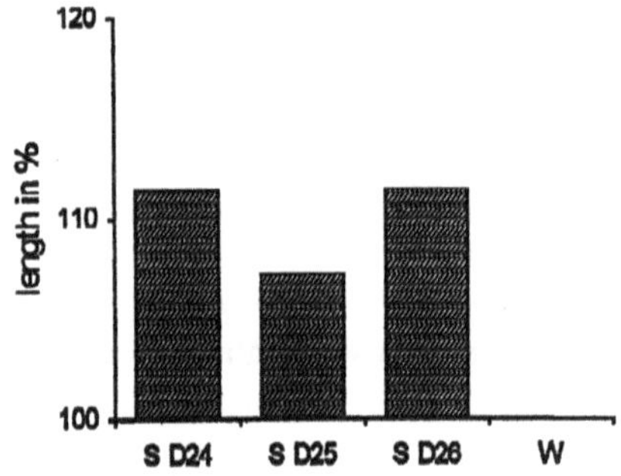

Fig. 1: The influence of the test dilutions silver nitrate D24, D25 and D26 in experiments that lasted 7 days. Ordinate: mean stalk length. 100% refers to the stalk length under the influence of unprepared solvent. Abscisses: S, silver nitrate. For further information, see text and table 5.

However, as can be seen in Tab. 1, the overall effect pattern in this study has in fact been caused by only about half of the single experiments (marked with *x* in Table 1).

SET I

Duration of growth 7 days

Exp. no. & N_{grains}	S D24	S D25	S D26	water
1: (4 x 120)	86.8 ± 27.9	84.2 ± 15.6	71.1 ± 40.5	31.4 ± 27.1
2: (4 x 160)	93.8 ± 33.6	69.3 ± 45.15	87.2 ± 23.2 x	41.3 ± 12.6
3: (4 x 120)	85.3 ± 25.9	68.4 ± 13.4	98.0 ± 15.3 x	54.6 ± 29.5
4: (4 x 160)	67.9 ± 17.2	57.5 ± 31.8	62.1 ± 31.5 x	56.8 ± 19.6
5: (4 x 200)	80.2 ± 27.8	76.3 ± 25.9	83.7 ± 28.2 x	59.8 ± 33.1
6: (4 x 120)	88.4 ± 24.8	86.3 ± 25.3	97.7 ± 13.6 x	67.8 ± 30.7
7: (4 x 120)	80.6 ± 26.8	73.9 ± 26.6	83.9 ± 22.6 x	69.6 ± 36.0
8: (4 x 200)	87.3 ± 28.1	80.6 ± 30.0	82.7 ± 30.1 x	74.4 ± 28.7
9: (4 x 180)	102.1 ± 34.8	98.0 ± 30.9	96.3 ± 41.6	74.7 ± 27.3
10: (4 x 120)	78.2 ± 15.7	87.4 ± 26.2	79.5 ± 18.4	75.8 ± 19.3
11: (4 x 240)	90.8 ± 29.2	94.9 ± 35.7	97.0 ± 30.6 x	92.3 ± 33.5
12: (4 x 200)	91.5 ± 28.5	99.0 ± 33.9	99.1 ± 31.5	94.8 ± 32.5
13: (4 x 200)	99.5 ± 38.0	105.3 ± 38.6	102.1 ± 36.6	96.4 ± 32.0
1 - 13: (4 x 2140)	91.3 ± 31.1	87.8 ± 33.4	91.3 ± 32.3	81.9 ± 35.0

Tab. 1: Stalk lengths in experiments on diluted and agitated substances that lasted 7 days. Left column: Number of experiment and number of grains for each of the 4 groups. Arithmetical mean (mm) + 1 SD for the groups treated with silver nitrate D24, D25, D26 and water, respectively. The experiments are arranged according to the stalk lengths in the control group. For further information, see text.

Interestingly, all experiments that showed the pattern discussed above were characterized by a stalk length of the control group between 40 and 75 mm.

SET II: Observation after 5 days, first experiments (W. Pongratz)

Experiments with 3 groups of 890 grains each, treated with silver nitrate D24, D25 and D26 and one group of 690 grains, treated with unprepared solvent, respectively, were performed.

Again, the values for the silver nitrate-groups are generally above the control-value (see Figure 2). When the test substances were mutually compared, both the group treated with the dilution D24 as well as with the dilution D26 showed more growth than the group treated with the dilution D25 (+ 8.7% and +10.6%). This difference is statistically significant ($P < 0.01$ and < 0.01).

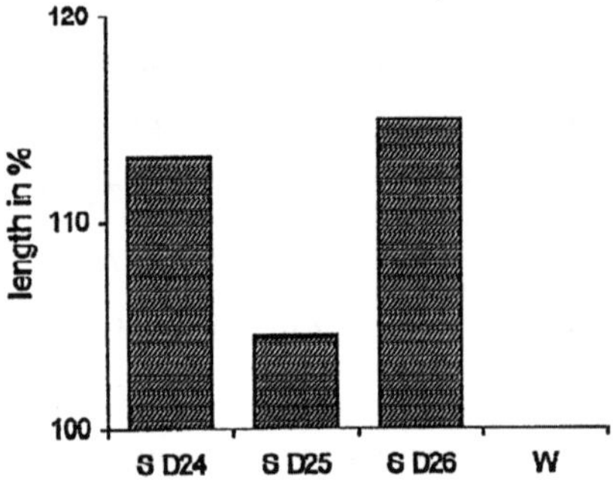

Fig. 2: The influence of the test dilutions in the first experiments that lasted 5 days. For further information, see Fig. 1.

All of the separate experiments of this study are characterized by a stalk length of the control group between 40 and 70 mm (see above) and all but one of them show the effect pattern 'stimulation - less stimulation - stimulation' as is presented in Table 2.

SET III: Observation after 5 days, control experiments (W. Pongratz, C. Endler, A. Nograsek)

Control experiments with a total of 4 groups of 510 grains each, treated with silver nitrate D24, D25 or D26 or with unprepared solvent, respectively, were performed in the laboratories of W. Pongratz, C. Endler and A. Nograsek (Table 3). When these data were pooled, again, the values for the silver nitrate - groups were generally above the control - value (Figure 3). When the test substances were mutually compared, both the group treated with the dilution D24 as well as with the dilution D26 showed more growth than the group treated with the dilution D25 (+ 16% and + 10%). This difference is statistically significant ($P < 0.01$ and < 0.01).

SET II

Duration of growth 5 days

Exp. no. & N_{grains}	S D24	S D25	S D26	water
14: (3 x 240)	09.1 ± 15.2	04.3 ± 08.5	10.0 ± 17.2	-
15: (3 x 60)	no separate protocol available			
16: (4 x 80)	84.6 ± 24.4	81.6 ± 22.3	86.5 ± 27.2	57.4 ± 31.9
17: (4 x 100)	83.8 ± 24.9	80.0 ± 23.5	86.0 ± 26.3	59.8 ± 33.1
18: (4 x 60)	84.4 ± 23.9	86.0 ± 23.0	82.7 ± 29.2	60.4 ± 34.2
19: (4 x 60)	86.8 ± 26.4	79.6 ± 20.6	88.9 ± 29.2	56.8 ± 32.6
20: (4 x 50)	82.1 ± 26.8	73.6 ± 21.4	92.8 ± 20.1	64.7 ± 33.6
21: (4 x 100)	83.2 ± 23.8	80.3 ± 24.0	88.4 ± 19.8	59.4 ± 32.0
22: (4 x 140)	30.6 ± 28.7	20.4 ± 26.3	29.7 ± 29.9	44.0 ± 37.4
14 - 22: (3 x 890, 1 x 690)	60.1	55.5	61.1	53.1

Tab. 2: Stalk lengths in the first experiments on diluted and agitated substances that lasted 5 days. For further information, see Tab. 1.

In the experiments of Pongratz, the typical "V"-form pattern is found, but only the difference between the groups treated with the dilutions D24 and D25 is statistically significant ($P < 0.01$).

In the experiments of Endler, the typical "V"-form pattern is found, but at the 1%-level, the difference is not statistically significant (D24 vs D25: $P < 0.05$).

In the experiments of Nograsek, the typical "V"-form pattern is found and both the difference between the groups treated with the dilutions D24 and D25 as well as between D25 and D26 is statistically significant ($P < 0.01$ and < 0.01).

SET III

Control experiments

Exp. no. & N_{grains}	S D24	S D25	S D26	water
Laboratory: W. Pongratz				
23: (4 x 200)	64.2 ± 23.8	55.6 ± 25.2	56.0 ± 27.3	47.8 ± 22.8
Laboratory: C. Endler				
24: (4 x 210)	33.0 ± 22.1	29.1 ± 20.7	31.3 ± 21.8	32.4 ± 21.4
Laboratory: A. Nograsek				
25: (4 x 100)	54.9 ± 20.4	45.5 ± 19.0	59.7 ± 19.6	42.7 ± 21.4

Tab. 3: Stalk lengths in control experiments on diluted and agitated substances that lasted 5 days. For further information, see Tab. 1

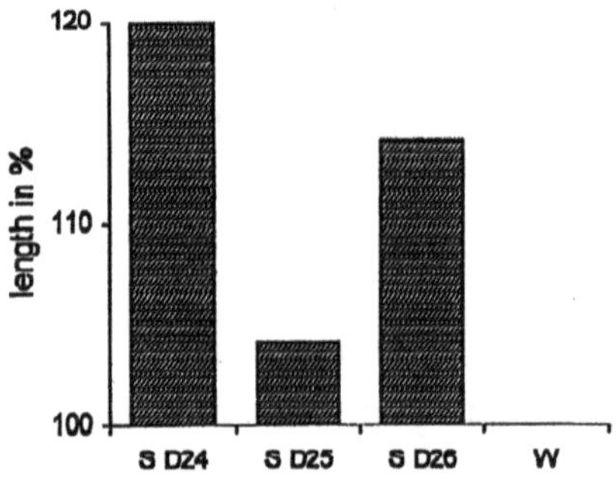

Fig. 3: The influence of the test dilutions in SET III. For further information, see Fig. 1.

Experiments SET IV on step by step agitated substances .
(C. Endler, W Pongratz)

Furthermore, experiments with 4 groups of 750 grains each, treated with silver nitrate D24/1, D24/2, D24/3 and unprepared solvent, respectively, were performed. The comparison of the mean stalk length again showed a characteristic effect pattern of the three test substances. The values for the silver nitrate - groups were generally above the control - value (see Figure 4). The difference is statistically significant. When the test substances were mutually compared, both the group treated with the dilution D24/1 as well as with D24/3 showed more growth than the group treated with the dilution D24/2

(+ 2.2% and + 9.6%). The difference between the groups treated with dilutions D24/1 and D24/2 is statistically significant ($P < 0.01$).

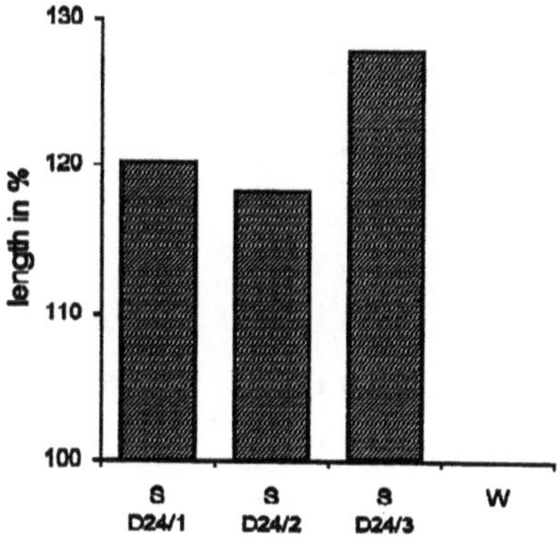

Fig. 4: The influence of the test dilutions silver nitrate D24/1, D24/2 and D24/3. For further information, see Fig. 1 and text.

Discussion

The study presented here is based on two previous findings: (a) that a concentrated solution of silver nitrate (10 mg per ml water), *inhibits* the germination of wheat seedlings (unpublished material) and (b) that this very solution, when diluted 10^{23}-fold in a special step by step process including 24 phases of vigorous agitation (called "D24") *stimulates* stalk growth. The latter was found in two previous multicenter blind studies (3,4). In the blind study presented in this paper, we have focused on the comparison of three different dilutions, namely silver nitrate D24, D25 and D26. In different series of experiments, it was found that adding one more step of dilution and agitation to D24 (= "D25") significantly *diminished* the effect of the test dilution, whereas adding two more steps (= "D26") again *enhanced* the stimulating effect. This was found by three independent researchers in three Austrian laboratories. The finding is also consistent with previous findings reported in literature (Kolisko 1926, Laupert 1995).

SET IV

Duration of growth 5 days

Exp. no. & N_{grains}	S D24/1	S D25/2	S D26/3	water
Laboratory: C. Endler				
A: (4 x 120)	31.4 ± 23.0	26.7 ± 23.2	28.6 ± 23.1	27.5 ± 21.1
Laboratory: W. Pongratz				
B: (4 x 120)	89.0 ± 35.0	86.6 ± 35.7	97.0 ± 45.8	61.4 ± 50.7
C: (4 x 120)	83.5 ±41.6	94.3 ± 34.8	108.6 ± 42.5	68.4 ± 32.3
D: (4 x 200)	78.6 ± 25.1	73.6 ± 31.4	76.3 ± 29.1	72.4 ± 24.4
E: (4 x 160)	69.8 ± 29.9	66.7 ± 29.2	70.0 ± 28.0	61.3 ± 42.3
A - E: (4 x 750)	69.2 ± 37.8	68.0 ± 39.1	73.5 ± 43.0	57.5 ± 38.7

Tab. 4: Stalk lengths in experiments on step by step agitated substances. For further information, see Table 1.

	SET I	SET II	SET III			SET IV	
	Pongratz	Pongratz	Pongratz	Endler	Nograsek		
24-25	++	++	++	+	++	24/1-24/2	-
24-26	-	-	++	-	-	24/1-24/3	++
24-water	++	++	++	-	++	24/1-water	++
25-26	++	++	-	-	++	24/2-24/3	++
25-water	++	-	++	+	-	24/2 water	++
26-water	++	+	++	-	++	24/3 water	++

Tab. 5: P - values in the experiments SET I - IV. ++, $P < 0.01$; +, $P < 0.05$, -, not significant (Fisher-test). For further information, see text.

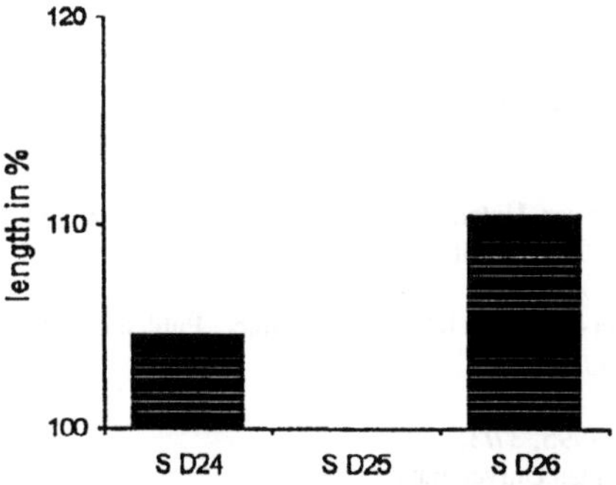

Fig. 5: The influence of the test dilutions D24, D25 and D26 (Kolisko 1926). 100% refers to the stalk length under the influence of silver nitrate D25. For further information, see Figure 1. (Adopted from Kolisko, modified.)

A "double switch" kinetic was postulated (D24: significant stimulation with regard to reference - D25: less or no stimulation - D26: again significant stimulation). This "V"-form effect pattern was investigated in experiments under different conditions of observation time. The pattern was less marked when growth was given a time of 7 days than when growth was given a time of 5 days. It was postulated that a final stalk length of 40 - 75 mm (water control group) was optimal to observe a clear effect. Further experiments with an observation time of 3 days and final stalk lengths of < 10 mm showed no difference between the groups treated with silver nitrate D24, D25 and D26 (unpublished material).

In the further blind experiments (duration 5 days), starting from a dilution silver nitrate D24, the process of step by step *dilution and agitation* was replaced by a process of only step by step *agitation*, alternating with resting times, without further dilution. In these experiments the pattern 'stimulation - less stimulation - stimulation' was found as above. This points to the conclusion that - in the range of the test substances D24 to D26 - it was not the process of step by step dilution, but of agitation, that caused the difference in the biological effect of the probes.

Further (preliminary) experiments were performed on a possible circadian specificity of the starting time of the experiments. Obviously, it is a conditio sine qua non in order to find the "V"-form effect pattern discussed here, that the experiments are started in the morning or noon, but not in the late afternoon (unpublished material).

At present, we have only very limited insight into the physical properties of such test dilutions (9-11 review in Endler and Schulte 1994) and into their physiological interconnections with the sensitive living system (Benveniste 1994, Endler et al. 1994(a,b), 1995, Youbicier et al. 1993, Endler and Schulte 1994). Further physiological and biophysical experiments have to be performed on the topic.

Acknowledgements

We would especially like to thank G. Kastberger, R. van Wijk, H. Walach, M. Moser and M. Haidvogl; as well as E. Lauppert who performed the experimental work in the laboratory of C.E.

References

Benveniste J et al. FASEB J 1994, 8/4, 398.

Demangeat JL et al. J Med Nucl Biophys 1992, 16/2, 135 .

Haidvogl M, Graz. Personal communication, 1997.

Kolisko L. Dornach: Goetheanum Verlag 1926.

Kratky K and Milavec T. Besteht ein qualitativer Unterschied zwischen Hochpotenz und reinem Lösungsmittel? In: Bergsmann O (ed). Wien: Facultas 1994.

Lauppert E. Thesis. University of Graz 1995.

Endler PC and Schulte J (eds), Dordrecht - Boston - London: Kluwer Academic Publishers 1994.

Endler PC et al., J Vet & Human Toxicol 1994a, 36/1, 56.

Endler PC et al., FASEB J 1994b, 8/4, 400.

Endler PC et al., J Vet & Human Toxicol 1995 1995, 37/3, 259.

Pongratz W et al.. In: Stacher A (ed). Wien: Facultas Universitätsverlag 1991, 385.

Pongratz W and Endler PC, in Endler PC and Schulte J (eds); Dordrecht - Boston - London: Kluwer Academic Publishers 1994.

Popp FA, in in Endler PC and Schulte J (eds); Dordrecht - Boston - London: Kluwer Academic Publishers 1994.

Smith CW. In: ZDN (ed). Essen: Verlag für Ganzheitsmedizin 1990; in in Endler PC and Schulte J (eds); Dordrecht - Boston - London: Kluwer Academic Publishers 1994.

Strasburger E. New York: G Fischer Verlag 1990

Weingärtner O. Berlin: Springer 1992.

Youbicier - Simo BJ, Boudard F, Mekaouche M et al., Int J. Immunotherapy 1993, IX/3, 169.

THE METAMORPHOSIS OF AMPHIBIANS AND INFORMATION OF THYROXIN STORAGE VIA THE BIPOLAR FLUID WATER AND ON A TECHNICAL DATA CARRIER; TRANSFERENCE VIA AN ELECTRONIC AMPLIFIER

PC Endler, C Heckmann. E Lauppert, W Pongratz, J Alex, D Dieterle, C Lukitsch, C Vinattieri, CW Smith, F Senekowitsch, H Moeller, J Schulte

Stimulating effect: It is known that the thyroid hormone thyroxin (tetraiodothyronine, T_4) has a decisive influence on the speed of metamorphosis in amphibians. If thyroxin is added to the water in a test basin until the final concentration in the basin is approx. 10^{-8}, then this induces and accelerates the laboratory animal's metamorphosis [1,2,3]. In previous experiments involving grass frog larvae (*Rana temporaria*) in the two-legged stage, L-thyroxin-sodiumpentahydrate at a concentration of 10^{-9} also accelerated the speed of metamorphosis in comparison to the control solutions (see methods) [4].

Inhibiting effect: However, thyroxin is able to do more than simply accelerate metamorphosis activity. At a thyroxin concentration of 10^{-7}, acceleration occurs at such a rate that malformations develop [5]. At a concentration level of 10^{-7}-10^{-6}, such thyroxin-induced malformations lead to a developmental block, or even to death, in Rana temporaria (unintentional results of earlier studies, see [4]).

1. Earlier work had provided indications that thyroxin at medium concentration levels may have an inhibiting effect, and that this effect may be varied by way of the preparation process, especially via gradual dilution and succussion of the trial substance [6]. However, experiments involving high dilutions suggest that this inhibition occurs only when the animals are in a certain starting position. When conducting experiments which progress relatively quickly thyroxin solutions at medium concentration levels

J. Schulte and P.C. Endler (eds.), Fundamental Research in Ultra High Dilutionand Homoeopathy, 155-187.

could also be expected to have a stimulating effect, analogous to the conditions which apply to high dilutions.

The experiments described later in this text (1.1) aimed to examine the effect of conventionally prepared thyroxin solutions at lower and medium dilution levels. The experiments described in 1.2 explored the question whether effects determined in 1.1 could be influenced using a special preparation process, particularly by way of gradual dilution and succussion. These experiments were conducted in view of the possible risks in the production of conventional hormone preparations [6], as well as of a certain production method used in homoeopathy [7,8, see also the article regarding research classification in homoeopathy in this book].

2. Further experiments dealt with the biophysical properties of the process of information transference. Since even dilutions which should theoretically no longer contain a single thyroxin molecule have a biological effect, it must also be possible to transfer "molecular bio-information" through water ([8-15], an overview can be found in [16]). The observed inhibition of young frogs' climbing activity, for example, can only be explained in this way [8]. It has been assumed that perimolecular energy fields play a role in the interactions between biologically active substances and the organism [further literature in 16, pp. 246].

In the experiments described in 2.1.1, the effect of the thyroxin dilution DH30 (concentration nominally 10^{-30}, test basin concentration after the first dose accordingly 10^{-35}), which was added to the basin water drop by drop, was examined. In the experiments described in 2.1.2, this thyroxin dilution was put into glass vials, which were subsequently sealed and hung into the basin water. In the experiments described in 2.2, an attempt was made to digitize the information contained in the thyroxin dilution.

3. In these tests, an attempt was made to transfer the information contained in molecular thyroxin dilutions by way of an electronic device (bio-resonance amplifier).

4. In these experiments, a thyroxin solution with an inhibiting effect on metamorphosis activity, and one with a stimulating effect, were administered together. The intention was to examine the "principle of similarity" known from homoeopathy as to its applicability in this case [7, and Research Classification in this volume].

5. The examinations described above were supplemented by additional experimental variations. Based on initial, rather coincidental indications, as well as on theoretical considerations (see Table 1), the question was explored, whether a stimulating effect could also be created with a thyroxin dilution which had shown an inhibiting effect in previous studies.

Note: This article focuses above all on experiments which were initially carried out in one of the Graz laboratories by W.P or C.E. and on account of their positive outcome were later repeated by other researchers in Graz or at other research centres. It does not include experimental variants which showed no significant effects of the test substances upon first trial in Graz. These experiments depart from those presented here only in terms of the time of season, and consequently the average speed of metamorphosis, or the elevation of the biotopes from which the animals were taken. For further details on this the reader is referred to [4].

OVERVIEW

The series of studies presented here was conducted during the years 1990-1997, mostly by several independent researchers and in several independent laboratories. Next to a critical examination of older data, it concerns the assumption that low energy (correlated, long-range) electromagnetic oscillations are a part of bio-information.

1. Lower and medium dilution levels

The effect of thyroxin solutions on the speed of transition from the two-legged to the four-legged stage in the grass frog *Rana temporaria* was examined. The solutions were added directly to the basin water at regular intervals. Starting from the first dose, the dilution was each time increased by the factor 10^5.

1.1 Conventional dilution of thyroxin

Reversal of the known hormonal effect?

A thyroxin solution, which was prepared in one step and without planned succussion, i.e. to a large extent merely with the aid of diffusion, and which, after the initial dose into the basin, was present at a concentration of 10^{-9} g/ml (10^{-9} M, from here onwards simply called 10^{-9}), had a stimulating effect on metamorphosis activity. This effect was no longer detectable at a concentration of 10^{-11ff}.

1.2 Special (homoeopathic) dilution process, low dilution level - reversal of the known hormonal effect?

In some, but not all of the experiments carried out, an inhibiting effect was found at a concentration of 10^{-11} and 10^{-13}, if the thyroxin solution had been prepared according to homoeopathic rules, that is using gradual steps of dilution and intermittent succussion. Certain components of the production process of conventional hormone preparations could unintentionally neutralize the preparations' desired effect, or even reverse it. On the other hand, such an effect reversal could be deliberately used therapeutically.

2. Non-molecular transference of hormonal information in homoeopathic high dilutions -clear reversal of the hormonal effect

2.1 Water as intermediate storage medium

In the experiments described in 2.1.1, the influence of high thyroxin dilutions on metamorphosis activity was examined. The solution added in drop form, and the corresponding water control, were also diluted in accordance with the procedure described above, i.e. in steps of 1:10, with subsequent succussion after each step. Based on the way it is produced and on its nominal concentration of 10^{-30}, this high-level thyroxin dilution is called DH30. After the initial dose was added to the basin, the nominal concentration stood at 10^{-35}. The thyroxin solution DH30 caused metamorphosis activity to slow down. In the experiments described in 2.1.2, the thyroxin solution DH30 and the corresponding water control were put into soft soda glass vials, which were subsequently sealed and hung into the basin water. In some, but not all experiments, this, too, caused a metamorphosis inhibition.

2.2 A data carrier as intermediate storage medium (CD)

In the experiments described in 2.2, vials containing the same solutions as in 2.1 were put into an input coil, which was connected to a filter and an amplifier. Frequencies were digitized, transferred onto CD in a multiplex procedure, and from there transferred onto an output spool, into which glass vials filled with water were put as "receivers". The trial substances thus produced were put directly into the basin water in regular intervals. In this first experiment, an inhibiting effect was found, comparable to that in 2.1.

3. Transference using electronic amplifiers (bio-resonance device)

In these experiments, glass vials containing an unsuccussed thyroxin one-step dilution of the (toxic, metamorphosis inhibiting) concentration 10^{-3} (10 mg for 10 ml ethanol, 1,25 mM, from here on called DD3), or water respectively, were placed into an input coil, one end of which was connected to a special amplifier (frequency linear from the DC region to the HF region) via an isolated individual wire. Glass vials containing water were placed into an output coil connected in the same way. The substances thus produced were put directly into the basin water at regular intervals. Again, an inhibiting effect was found.

4. Curative effect in cases of hormonal hyperstimulation

A combination of the unsuccussed one-step dilution thyroxin 10^{-4} (from here on called DD4) and a thyroxin solution 10^{-8} (DH8), prepared according to homoeopathic rules, was examined in a direct pipetting procedure, as had been done in 2.1. Hence, after the initial dose, both solutions had been diluted by the factor 10^{-5}. In this explorative experiment the stimulating effect of Thyroxin DD4 (see 1.1) was reversed by simultaneously adding a dose of DH8.

This finding is interpreted with the "similarity rule" in mind, one of homeopathy's fundamental principles (see the article regarding the classification of pure research in homoeopathy in this book).

5. Symptom reduction and symptom increase as bivalent possibilities of one and the same aspect of hormonal information

With the continuation of treatment as in study 2.1.2 (constant presence of thyroxin DH30 in glass vials) beyond the juvenile stage, a clear acceleration of the water-land transition could be observed when the juvenile frogs left the basin water, after an initial inhibition of the transition to the four-legged stage. On the other hand, an inhibiting effect was observed with regards to climbing activity, when as yet untreated juveniles were treated with thyroxin DH30 on a short-term basis (minutes). The results of study 5 are discussed to the effect that, next to the animals' starting position, it is the frequency at which signals are given, which decides whether the information of thyroxin DH30 leads to inhibition or stimulation.

METHODS

1. Study concerning the reversal of the known hormonal effect at the medium dilution level

Origin and developmental stage of the animals. The main experiments were carried out with Rana temporaria from various Austrian ponds (altitude 400m – low site - and 1500m above sea level – highland -, see results section) in the summer. Additional experiments were carried out with Bufo bufo. For the experiments only tadpoles in a two-legged stage were used. This stage is defined by the degree to which the hind legs, which are little developed at this point, are spread. Ideally, the spread is such that it is just possible to look through the triangle formed by the tail and the upper and lower thighs, and it thus corresponds to the range of Gosner's stage 31 [17]. In the different experiments, the tadpoles were observed all in all 10 - 20 times in eight-hourly intervals. During this observational period, it only took a few minutes each time for the pre-formed forelegs to break through their skin pockets. Due to its temporal discreteness, the attainment of the four-legged stage appears to be a very good parameter.

Other experimental conditions. The animals' development was observed in disposable synthetic basins. The room temperature normally was 23+/-1-C. Indirect daylight was predominantly used as a source of light, and blanched lettuce leaves as food ad libitum. Each basin contained 300 ml of tap water per animal. The number of animals per basin was the same in each experiment (usually 10 – 20).

Preparation of the solution and its administration to the laboratory animals. The tadpoles were exposed to the following solutions.

1.1 Conventional dilution process

Preparation of one-step thyroxin dilutions by diffusion. These solutions were prepared by slowly dissolving L-tetraiodothyronine-sodiumpentahydrate (T_4) (careful turning of the containers, diffusion) in mixtures of ethanol and bi-distilled water: 10 mg T_4 in 100 ml 40% ethanol for DD4, which thus stood at 10^{-9} T_4 after a single dose was added to the basin water; 1 mg T4 in 100 ml 4% ethanol for DD5 corresponding to 10^{-10} T_4; and 0.5 mg T4 in 500 ml 0.4% ethanol for DD6, leading to a concentration of 10^{-11} T_4. Analogue ethanol-water mixtures were used as *controls*.

Preparation of thyroxin dilutions by gradual pipetting. In the preparation of these solutions, 10 mg of T_4 was dissolved in 100 ml of 40% ethanol (intermediate product: 10^{-4} T_4), by slowly turning the bottle (medicine bottle with screw-on lid) at a temperature of 35 °C. Using disposable material, 1 ml of this was then pipetted into a sterile, 20 ml medicine bottle containing 9 ml of bi-distilled water (dilution 1:10). The thyroxin dilution thus produced was called DP5. From it, the thyroxin dilutions DP6 ff were produced in an analogous fashion. Normally, analogue ethanol-water mixtures were used as *controls*.

1.2 Special ("homoeopathic") dilution process

Gradually diluted and strongly successed thyroxin. In the preparation of these solutions, the same process as above was used (pipetting). The derived dilution of 10^{-5} T_4 was then successed according to a standardized procedure [8]:

in short, regular intervals, and with the lid tightly screwed on, the medicine bottle's bottom is struck 30 times against a solid hard rubber base, in order to create a mechanical shock effect. Afterwards it is left standing for one minute (rest period). In principle, this procedure is in accordance with the rules of homoeopathic practice. The thyroxin dilution produced in this way was called DH5. From it, the thyroxin dilutions DH6 ff were produced in an analogous fashion.

Control solutions. In different experimental blocks, a.) unsuccessed water and analogue water-alcohol mixtures (b. unsuccessed, c. successed – WDH –) were used as control solutions (see results).

All DD, DP, and DH thyroxin solutions and their matched water controls were sampled several times for analysis with a total T_4 assay (measuring range 4.10^{-9} to $2.4.10^{-7}$ gT4/ml). Except for occasional threshold-level reactions in the water controls and a single overshooting value in one of the thyroxin dilutions the measured values never deviated from the nominal concentration by more than a factor of 2 [45].

Administration of the solution. In eight-hourly intervals, 3ml of test solution (thyroxin solution and control solution) per animal and 300 ml of water was altogether pipetted 20 times into the basin. This means that DH6, for instance, was 10^5 times diluted after the first dose had been added, bringing it to a T_4-concentration of 10^{-11}, while the tenth dose raised it to 10^{-10}.

Ascertaining and evaluating the data. As the duration of the experiments differed according to the time of season (range 3 - 10 days) it was decided to normalize the obtained data with respect to time. This was done on the assumption that differences in speed of metamorphosis attributable to thyroxine treatment would override differences in duration between experiments. Following a suggestion by R. Lüdtke, Institut für Medizinische Informationsverarbeitung Tübingen University, the points in time when the 10th, 20th, 30th etc. percentile of water control animals had reached the 4-legged stage were defined as reference points. The time interval between reference points ranged 6 - 26 hours. In this way it was possible to aggregate for each of the reference points the cumulative frequency of animals treated with a specific solution having reached the 4-legged stage. Aggregate values obtained at the reference points for each of the types of treatment were analysed by chi-square tests using 4-field Tables with aggregate frequencies of 2- or 3-legged animals as complement. Unless stated otherwise the p-values quoted in the text refer to chi square tests. In addition, data of studies with homoeopathically prepared dilutions DH6, DH8 and DH30, added directly to the basin water, were compared by means of survival analysis.

Note: as a consequence of the above mentioned method of ascertaining the data, p-values may eventually be different from those calculated for previous publications.

Laboratory, researchers and coding. For details on laboratories and researchers, see the results section. The full names and affiliations of the researchers are to be found in the acknowledgments section. All experiments were carried out blindly.

2. Non-molecular transfer of the high dilution's hormonal information

2.1 Water as intermediate storage medium for hormonal information

2.1.1 Use of a successed solution added to the basin water in drop form

Animals as is described in 1) were used. For details, see results section.

Other experimental conditions. During the main experiments, room temperature stood at 19+/-1 ∞ C. Further details are described in 1.

Preparation of the solutions and their administration to the laboratory animals. The principle of the preparation of the thyroxin dilution DH30, and of the analogue water control, has been described in 1.2. In order to exclude the possibility of hormone-specific pollutions, we took the precautions described in [4, p. 46]. In 48-hourly intervals, doses of 2-3 drops were added respectively to the 51 and 81 test basins containing tap water. The effect of the gradually diluted, strongly successed thyroxin solution DH30 on the tadpoles was observed as described in 1.

Ascertaining and evaluating the data. This was done as in 1. The development towards the juvenile stage was followed by determining the corresponding cumulative frequencies (*fb*). Standard deviations were found using a process of variance analysis.

For *experimental laboratories and researchers*, see results section.

2.1.2 Use of a succussed solution contained in sealed glass vials

Administration of the solutions to the laboratory animals. The solutions (DH30) were administered to the animals as follows: 8ml of the test- or the control solution were poured into soft soda glass vials with an optic transmission spectrum >350 nm, which were subsequently sealed. The coded vials were hung into the respective basins. In one experiment (C.L.), thyroxine *DH6* was compared to control. The use of substances in sealed glass vials for scientific and therapeutic purposes has been tried on several occasions (van Wijk and Wiegant, Smith, Flyborg, Hoffmann, cited in [16, p.48]). *For further details, see 1. and results section.*

2.2 A data carrier as an intermediate storage medium

Preparation of the solutions and their administration to the animals. Brown glass bottles (medicine bottles) containing thyroxin solution DH30 or control water DH30 respectively were placed into an input coil connected to a filter and an amplifier with an amplification of 10^6. Frequencies in the Hz and the kHz regions were digitized via the Nyquist frequency, buffered in a RAM, and transferred onto CD using a multiplex procedure. The noise-muted, filtered signal was brought back to the original analogue level by the factor 10^{-6}. During the playback process, only the region between 20 Hz and 20 kHz was taken into account. The glass bottles were placed into the output coil for 4 minutes (manufacturer of the utilized device: Subwave, Austria). Following this process, the original substances were again strongly succussed (30 strikes). Of the trial substances thus produced, 3 drops were added to the basin water after light succussion in 48-hourly intervals. *For further details, see 1. and results section.*

3. Transference via bio-resonance device

Preparation of the trial substances and their administration to the animals. After repeated light succussion, brown glass bottles containing a 1 mM thyroxin solution (DD3), or control water respectively, were placed into an input cup connected to a special amplifier (linear from DC to HF; manufacturer: Regumed and Med-Tronik, both FRG) via an isolated individual wire. Brown glass bottles containing water (200 ml in a 220 ml volume) were placed into an output beaker, which had been connected in the same way, for 4 minutes (device built by Regumed) and 15 minutes (device built by Med-Tronik) respectively. The options "A" (no phase reversal) and "Amplification 40", offered by the manufacturers, were selected. Following this process, the substances in the output bottles were struck 30 times in short, regular intervals against a base, as in the process of "succussion" (method 1.2). For each of the trial substances thus produced in the output spool, 8 ml of basin water were replaced by 8 ml of trial substance in eight-hourly intervals. Before being added, the substances were again lightly succussed. Explanations concerning the standardization of this method can be found in [20-25]. *For further details, see 1. and results section.*

4. Curative effect in cases of hormonal hyperstimulation

In these experiments the combined administration of the thyroxin solutions DD4 (concentration: 10^{-4}, basin water concentration after the initial dose: 10^{-9}, after the tenth: 10^{-8}) and DH8 (10^{-8}, final basin concentration: 10^{-13} or 10^{-12} respectively) was compared to the combined administration of thyroxin DD4 and water. In addition, a water control group was used. In eight-hourly intervals, all in all 20 doses of 3ml of trial substance (or 6ml, under combined administration) were added in drop form per animal and 300 ml of basin water. *For further details, see 1. and results section.*

5. Symptom reduction and symptom intensification as bivalent possibilities of one and the same aspect of hormonal information

The target criterion of these experiments was, next to the attainment of the four-legged and the juvenile stages, the animals' water-land transition. The respective treatments (see results section) were continued beyond the four-legged and the juvenile stages, until the animals had climbed out of the water of suitable aquaterraria. In the main study presented here, the treatment applied in study 2.1.2 (constant presence of thyroxin DH30 in glass vials) was continued until the target criterion of water-land transitions had also been attained.

Ascertaining and evaluating the data. The animals were first all put back into the water. Then, the number of animals climbing out of the water was monitored after 1, 2 and 3 minutes. Details of the monitoring method have been described in [8]. Aggregate values obtained at the reference points for each of the types of treatment were analysed by chi-square tests using 4-field Tables with aggregate frequencies of animals in the water as complement.

For further details, see 1. and results section.

RESULTS

1. Study concerning the reversal of the known hormonal effect at medium dilution levels

1.1 Conventional dilution process

Thyroxin solutions produced by pure diffusion. In these experiments carried out by C.E. et al. in Graz and by C.H. in Tubingen with an overall number of 170 – 200 animals per group according to the treatment, the effects of the one-step thyroxin dilutions DD4, DD5 and DD6 (no repeated pipetting, no succussion, see methods) were compared to those of the corresponding water controls. For DD4 (concentration after a single dose: 10^{-9}), the *fa*-levels for four-legged animals in the course of treatment stood at 10-30% above those of the respective water controls. These differences are statistically significant ($p<0.01$). The overall *fa*-levels for DD5 or DD6 (10^{-10} or 10^{-11}, after the first dose) were practically not different to those of the water control.

Thyroxin solutions produced by gradual pipetting. Also with thyroxine dilution DP5, DP6 or DP8, there was practically no difference to the control.

1.2 Special ("homoeopathic") dilution process

Thyroxin solutions produced by gradual pipetting and succussion.

DH4. In these experiments carried out by C.H. in Tübingen with a total of 120 animals per group, the effect of the successed thyroxin dilution DH4 was compared to that of the corresponding water. For the test substance (concentration after a single dose: 10^{-9}), the fa-levels for four-legged animals in the course of treatment stood at 20-40% above those of the respective water controls. This difference is statistically significant ($p<0.01$).

DH6 and DH8. The effect of the dilutions DH6 and DH8 were compared to water controls. Experiments were carried out by different researchers in different laboratories. E.L., W.P. and C.H. performed experiments (blocks have a different meaning here than experiments) in Graz; J.A. and C.H. in Tübingen, and C.L. in Vienna [26; 45-47]. A total of 2300 animals was used to test thyroxine DH6 versus water control and a total of about 1400 animals to test DH8 versus control. As can be seen in Figure 1 and Table 1, in 6 out 10 different experiments the fa-values of the DH8 series were lower for the thyroxin dilution than for the matched control, three experiments showed practically no difference, and in one there was an opposite trend. Pooling the values for all DH6 – experiments leaves practically no difference between the two groups, save for a weak inhibition (trend) in the TDH- group relative to water control towards the end of the experiment. Pooling all DH8 blocks reveals a somewhat more distinct inhibition by thyroxin relative to water control towards the end of the experiment (trend). An overall weakly significant inhibition ($p<0.05$) towards the end of the experiment is found only when the results from the medium dilution steps DH6 and DH8 are pooled. Data are not significant, when survival analysis is used. fb-levels were not recorded in these experiments. Data on further test- and control- dilutions are presented in [45,46].

DH6

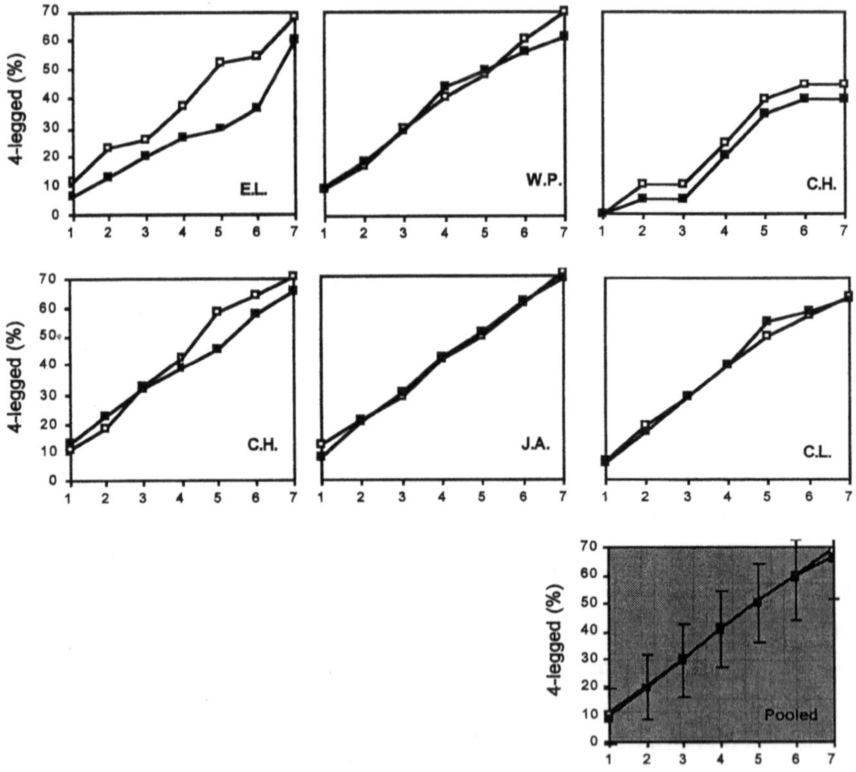

Figure 1: see figure caption next page.

DH8

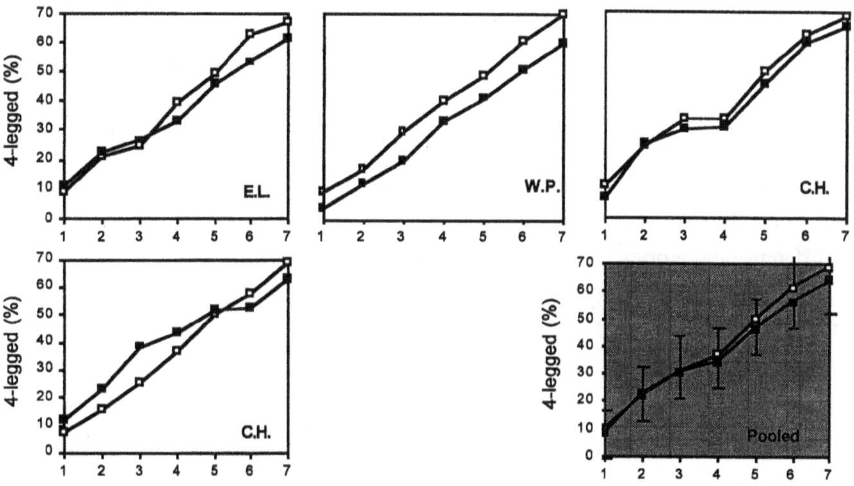

Figure 1: The effect of gradually diluted, succussed thyroxin solutions 10^{-11} and 10^{-13} on the speed of the metamorphosis from the 2- to the 4-legged stage in tadpoles. Ordinate: cumulative number of animals reaching the four-legged stage (*fa*, in %); abscissa: course of development, see methods. White squares: *fa*-levels of the water controls; black squares: *fa*-levels of animals treated with the respective test substances. TDH6, TDH8: gradually diluted, succussed thyroxin solutions, added dropwise to the basin water. WDH: homoeopathically prepared control. W: pooled data of homoeopathically and not homoeopathically prepared water controls. Absolute animal numbers are shown in Table 3.

2. Non-molecular transference of a highly diluted solution's hormonal information

2.1 Water as an intermediate information storage medium

2.1.1 Use of a succussed solution added to the basin water in drop form

DH30. The effect of the thyroxin dilution DH30 was compared to a water control. Experiments were carried out by different researchers in different laboratories: C.E. and W.P. performed experiments in Graz and R.v.W. in Utrecht [28]. A total of almost 1900 animals was used. As can be seen in Figure 2 and Table 2, in all three cases the fa-values of the TDH30 group were lower than their matched control values. This effect is retained upon pooling ($p<0.01$). Data are also significant when survival analysis is used ($p<0.01$). Also the *fb*-levels (observed in Graz and in Utrecht) were significantly below those of the control groups [4].

2.1.2 Use of a succussed solution contained in sealed glass vials

"DH30V / DH6V". The dilutions DH30 or DH6 (see Figure 3), sealed in glass vials, were compared to water controls. Experiments were carried out by different researchers in different laboratories as is shown in Figure 1 and Table 3. *DH30V:* C.E., M.G., W.P., D.D. performed experiments in Graz; C.V. in Turino and H.H. in Vienna [26; 47,48]. A total of about 3400 animals was used. As can be seen from Figure 3 and Table 3 five out of altogether 6 experiments on DH30V showed lower fa-values for the test than for the control group; one (very large experiment) showed practically no difference. Pooling all DH30V experiments revealed a significant inhibition by thyroxin DH30V relative to matched water control ($p<0.01$).

As the DH6V curve in Figure 3 shows (C.L. in Vienna), this experiment also yielded lower *fa*-values for the test than for the control group (trend).

In some, but not all experiments (Graz), also the *fb*-values were monitored. There was practically no difference between the test- and the control values.

2.2 A data carrier as intermediate storage medium

"DH30CD". Here, it was aimed to store information from thyroxin dilution DH30 or from water DH30 as control on data carrier and to replay it on water that was added to the basins. Experiments were carried out by one researcher (W.P.) in Graz, involving almost 500 animals. As can be seen from Figure 4, this experiment resulted in lower fa-values for the test than for the control group ($p<0.01$). Also the *fb*-levels were significantly below those of the control groups [44].

3. Transference via electronic amplifier

"DD3A". Here, it was aimed to transfer information from thyroxin DD3 or from water as control via an electronic amplifier. Experiments were carried out by two researchers: W.P. in Graz and C.V. in Torino, involving altogether more than 900 animals. As can be seen in Figure 5 and Table 5 in both experiments the fa-values for the test group were lower than in the control group. Pooling these figures led to an analogous result ($p<0.01$). Also the *fb*-levels were significantly below those of the control groups [44].

E.L.	Graz 1 ll r		
	W	TDH6	
	(160)	(30)	
1	18	2	-
2	36	4	-
3	42	6	-
4	60	8	-
5	84	9	**
6	88	11	-
7	109	18	-
8	115	19	-

W.P.	Graz 2 ll r		
	W	TDH6	
	(160)	(80)	
1	15	8	-
2	28	15	-
3	48	23	-
4	65	35	-
5	78	40	-
6	97	45	-
7	112	49	-
8	127	58	-

C.H	Graz 1 ll r		
	WDH6	TDH6	
	(20)	(20)	
1	0	0	-
2	2	1	-
3	2	1	-
4	5	4	-
5	8	7	-
6	9	8	-
7	9	8	-
8	9	8	-

C.H	Tubingen 1 ll r		
	WDH6	TDH6	
	(120)	(120)	
1	13	16	-
2	22	27	-
3	39	38	-
4	51	47	-
5	70	55	-
6	77	69	-
7	85	79	-
8	92	90	-

J.A.	Tubingen 2 ll r		
	WDH6	TDH6	
	(600)	(600)	
1	72	50	*
2	125	121	-
3	176	184	-
4	251	254	-
5	299	305	-
6	365	370	-
7	432	418	-
8	432	418	-

C.L.	Vienna ll r		
	W	TDH6	
	(300)	(300)	
1	20	18	-
2	57	51	-
3	87	88	-
4	119	119	-
5	149	165	-
6	172	175	-
7	190	187	-
8	251	238	-

	Pooled data		
	W	TDH6	
	(1150)	(1150)	
1	116	94	-
2	227	219	-
3	336	340	-
4	470	467	-
5	581	581	-
6	688	678	-
7	792	759	-
8	869	831	-

Table 1: Onset of the four-legged stage in the experiments represented in Figure 1, expressed in absolute numbers. ll r: Rana temporaria from a low site; hl r: from the highland biotope. -: $p>0.05$; *: $p<0.05$; **: $p<0.01$ (differences between test group and control group, chi-square test). Note: When, as in two out of altogether 6 cases, the test and matched control group within an experiment were of different size, then the larger group was accorded a correspondingly smaller weight in the overall pooling. For example, if N was 80 for the tests group and 160 for the control group, then all *fa* –values of the control group were divided by 2 so that the subsequent pooling would be applied to groups of equal weight.

E.L.	Graz 1 ll r		
	W	TDH8	
	109	177	
1	10	20	-
2	23	39	-
3	27	47	-
4	43	58	-
5	54	81	-
6	68	94	-
7	73	109	-
8	84	126	-

W.P	Graz 2 ll r		
	W	TDH8	
	160	80	
1	15	3	-
2	28	10	-
3	48	16	-
4	65	27	-
5	78	33	-
6	97	41	-
7	112	48	-
8	127	56	-

C.H	Graz 1 ll r		
	W	TDH8	
	320	320	
1	37	23	-
2	79	80	-
3	107	97	-
4	109	98	-
5	158	145	-
6	199	190	-
7	219	208	-
8	253	253	-

C.H	Tubingen 1 ll r		
	WDH8	TDH8	
	120	120	
1	8	14	-
2	19	28	-
3	31	46	*
4	44	52	-
5	60	62	-
6	69	63	-
7	83	76	-
8	91	92	-

	Pooled data		
	W	TDH8	
	629	629	
1	63	52	-
2	135	142	-
3	189	188	-
4	229	213	-
5	311	290	-
6	385	352	-
7	431	399	-
8	491	478	-

Table 1 continued.

In summary, in most, but not all experiments, the following treatment forms lead to a delay (trend or significant effect) in metamorphosis in comparison to the animals correspondingly treated with water: 8-hourly addition of thyroxin dilution DH6 or DH8 in drop form (poor effect when the respective data are pooled); 48-hourly addition of thyroxin dilution DH30 in drop form; treatment with a thyroxin dilution DH30 sealed inside of glass vials by hanging the vials into the basin water (small effect when the data are pooled); addition in drop form of water, which had previously been exposed to digitized information of the thyroxin dilution DH30 stored on CD, into the basin water (only one set of experiments performed); addition of water, which had previously been exposed to electronically transferred information of the thyroxin dilution DD3, into the basin water.

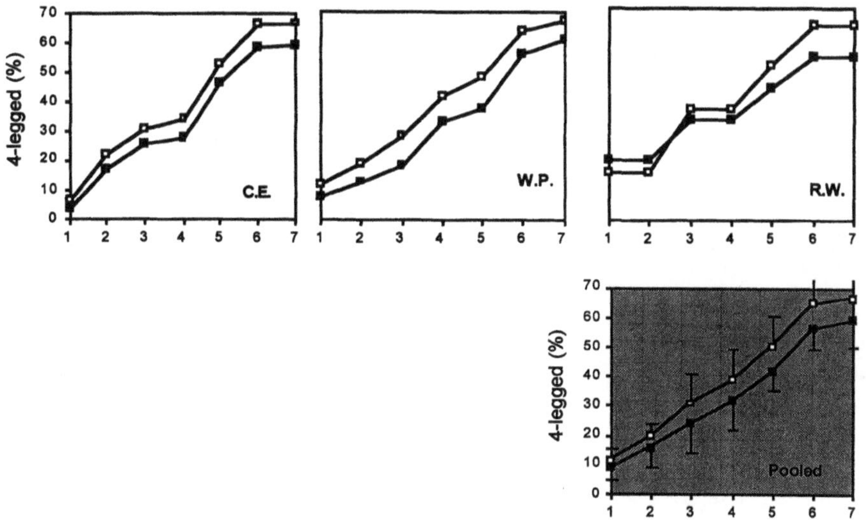

Figure 2: The effect of successed thyroxin solution 10^{-35}. DH30: gradually diluted, successed thyroxin solutions, added dropwise to the basin water. Absolute animal numbers are shown in Table 2.

C.E.	Graz 1 hl r		
	WDH30	TDH30	
	273	276	
1	17	10	-
2	60	48	-
3	84	70	-
4	94	76	-
5	145	129	-
6	181	161	-
7	182	164	-
8	227	211	-

W.P.	Graz 2 hl r		
	WDH30	TDH30	
	475	475	
1	58	36	*
2	91	61	**
3	132	89	**
4	198	159	**
5	226	177	**
6	299	264	*
7	319	287	*
8	403	352	**

R.W.	Utrecht 1 hl r		
	WDH30	TDH30	
	180	180	
1	29	36	-
2	29	36	-
3	66	60	-
4	66	60	-
5	92	78	-
6	116	98	-
7	116	98	-
8	153	151	-

Pooled data			
	WDH30	TDH30	
	928	928	
1	104	82	-
2	180	144	*
3	282	218	**
4	358	294	**
5	463	383	**
6	596	521	**
7	617	547	**
8	783	712	**

Table 2: Onset of the four-legged stage in the experiments represented in Figure 1.

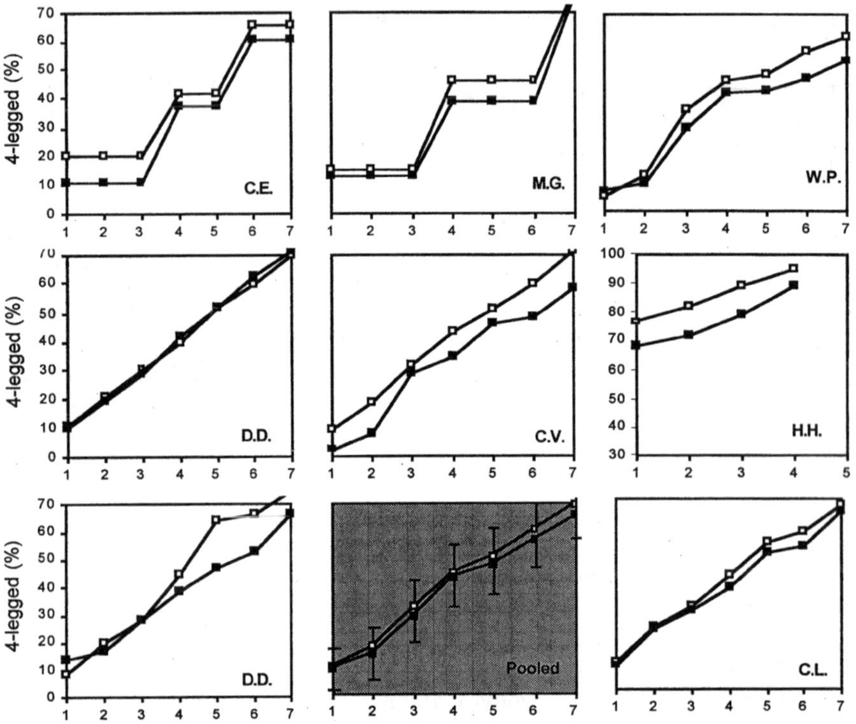

Figure 3: The effect of successed thyroxin solution 10^{-35}, sealed in glass vials. DH30V: gradually diluted, successed thyroxin solution, sealed in glass vial and hung into the water basin. Absolute animal numbers are shown in Table 3.

4. Curative effect in cases of hormonal hyperstimulation

A combined administration of the thyroxin dilution DD4 (which, when administered on its own, has a stimulating effect, see 1.1) and the thyroxin dilution DH8 (which, on its own, has an inhibiting effect, see 1.2) was compared to the combined administration of DD4 and water, and to a water control. This first experiment was carried out by C.H. in C.E.'s laboratory, using 20 animals in each of the three groups.The combined addition of DD4 and water caused an acceleration of metamorphosis activity in reference to the water control, while the combined addition of DD4 and DH8 caused metamorphosis activity *to slow down*. The difference in the *fa*-levels of both groups was 20-25% ($p<0.01$) at the same sampling points [26].

C.E.	Graz 1 ll b		
	WDH30V	TDH30V	
	180	180	
1	36	20	*
2	36	20	*
3	36	20	*
4	75	67	-
5	75	67	-
6	118	109	-
7	118	109	-
8	159	153	-

M.G.	Graz 1 hl r		
	WDH30V	TDH30V	
	126	126	
1	19	16	-
2	19	16	-
3	19	16	-
4	58	49	-
5	58	49	-
6	58	49	-
7	99	96	-
8	99	96	-

W.P.	Graz 2 ll r		
	WDH30V	TDH30V	
	612	612	
1	32	45	-
2	76	62	-
3	221	183	*
4	281	258	-
5	297	260	*
6	346	288	**
7	377	325	**
8	451	413	*

D.D.	Graz 2 ll r		
	WDH30V	TDH30V	
	612	612	
1	68	62	-
2	128	121	-
3	184	178	-
4	242	255	-
5	316	318	-
6	365	382	-
7	428	438	-
8	480	496	-

C.V.	Turin ll r		
	WDH30V	TDH30V	
	72	72	
1	7	2	-
2	14	6	-
3	23	21	-
4	31	25	-
5	37	33	-
6	43	35	-
7	51	42	-
8	55	47	-

H.H.	Vienna 2 hl r		
	WDH30V	TDH30V	
	108	108	
1	4	6	-
2	9	15	-
3	47	53	-
4	50	58	-
5	50	58	-
6	69	73	-
7	83	74	-
8	89	78	-

D.D.	Graz 2 ll r (Quartz)		
	WDH30V	TDH30V	
	36	36	
1	3	5	-
2	7	6	-
3	10	10	-
4	16	14	-
5	23	17	-
6	24	19	-
7	27	24	-
8	28	28	-

	Pooled data		
	WDH30V	TDH30V	
	1746	1746	
1	173	159	-
2	295	251	*
3	551	491	*
4	769	741	-
5	874	819	-
6	1045	975	*
7	1208	1131	**
8	1390	1339	*

C.L.	Vienna 2 hl r		
	WDH6V	TDH6V	
	300	300	
1	32	27	-
2	68	65	-
3	93	86	-
4	127	111	-
5	161	150	-
6	173	157	-
7	202	196	-
8	232	229	-

Table 3: Onset of the four-legged stage in the experiments represented in Figure 3. ll b: Bufo bufo from a low site. Quartz: in this experiment, quartz glass vials were used instead of soft soda glass vials. For further information, see Figure 3. Note: One of these experiments (C.E.) was carried out with Bufo bufo. They have been included in the pooling to give a preliminary overview.

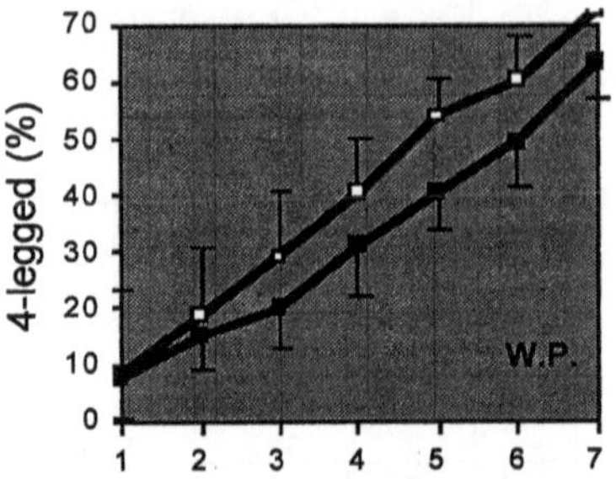

W.P Graz 2 ll r			
	WDH30 CD	TDH30 CD	
	234	234	
1	20	19	-
2	44	36	-
3	68	47	*
4	95	73	*
5	126	95	**
6	141	115	*
7	169	148	*
8	183	160	*

Figure and Table 4: The effect of information from thyroxin, stored on a compact disk (DH30CD). Absolute animal numbers are shown in Table 4.

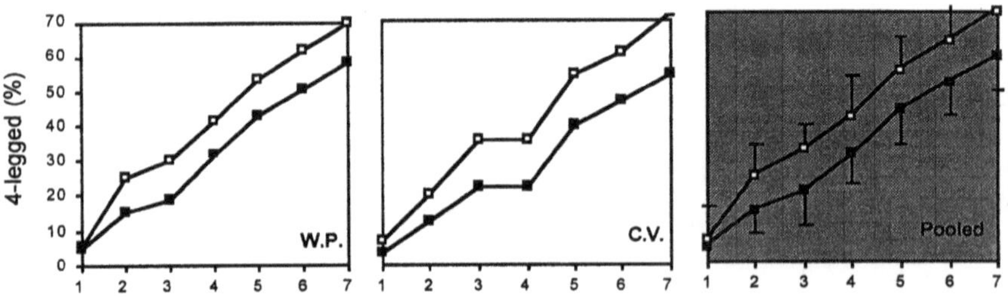

Figure 5: The effect of information from thyroxin, transferred via an electronic amplifier. (DD3A). Absolute animal numbers are shown in Table 5.

W.P.	Graz 2 II r		
	WDD3A	TDD3A	
	378	378	
1	21	17	-
2	95	57	**
3	114	71	**
4	158	120	**
5	203	162	**
6	234	192	**
7	262	220	**
8	291	244	**

C.V.	Turin II r		
	WDD3A	TDD3A	
	90	90	
1	6	3	-
2	18	11	-
3	32	20	*
4	32	20	*
5	49	36	-
6	55	42	-
7	65	49	*
8	70	50	**

	Pooled data		
	WDD3A	TDD3A	
	468	468	
1	27	20	-
2	113	68	**
3	146	91	**
4	190	140	**
5	252	198	**
6	289	234	**
7	327	269	**
8	361	294	**

Table 5.: DD3A

Table 5: Onset of the four-legged stage in the experiments represented in Figure 5. For further information, see there. The complete set of raw data has been published in [44].

5. Symptom reduction and symptom intensification as bivalent possibilities of one and the same aspect of hormonal information

This section summarizes a series of studies. When the test substance thyroxine DH30 was added at intervals of 8 hours, an *acceleration* was achieved as a result of experiments carried out on highland animals in autumn, when the dose interval in protocol 2.1.1 was reduced to 1/6. In fact, both the transition from the two-legged to the four-legged stage as well as that from the four-legged to the land stage [4, p.57] were accelerated.

Furthermore, in Graz, C.E. et al. performed experiments with 360 highland animals treated with thyroxin DH30 and 360 animals treated with water DH30, added at intervals of 8 hours. Again as above, the thyroxin *fa*-level was *above* the control level (5 to 10%, $p<0.05$). CE et al. also performed several experiments with lowland animals, with unclear results ($p>0.05$). In Tubingen, C.H. performed experiments with 120 lowland animals per group. Again as above, the thyroxin *fa*-level was *above* the control level (5 to 10%, $p>0.05$).

However, in the absence of an interval reduction (continuation of study 2.1.1), neither the transition from the 2 to 4-legged stage nor that from water to land was stimulated, the inhibiting effect prevailing instead in both cases [4, p.56]. The inhibiting effect was also observed with regard to climbing activity (water-land), when as yet untreated juveniles were treated with thyroxin DH30 on a short-term basis (minutes), irrespective of whether the test substance was added directly in drop form (C.E., W.P., [8], significant results), or locked into vials (C.E., W.P., M.G., F.W., [4, p.45-48], trends).

With the continuation of treatment as in study 2.1.2 (constant presence of thyroxin DH30 in glass vials) beyond the juvenile stage, a clear acceleration of the metamorphosis leading to the land stage could be observed upon leaving the basin water, after an initial inhibition of metamorphosis [4, p.58].

Experiments were carried out by different researchers: C.E., W.P. and M.G. performed experiments in Graz and H.H. performed experiments in Vienna. As can be seen in Figure 6 all four experiments yielded higher fa-values for the thyroxin DH30 group than for water control. Pooling the *fa*-values led to analogous result ($p<0.01$).

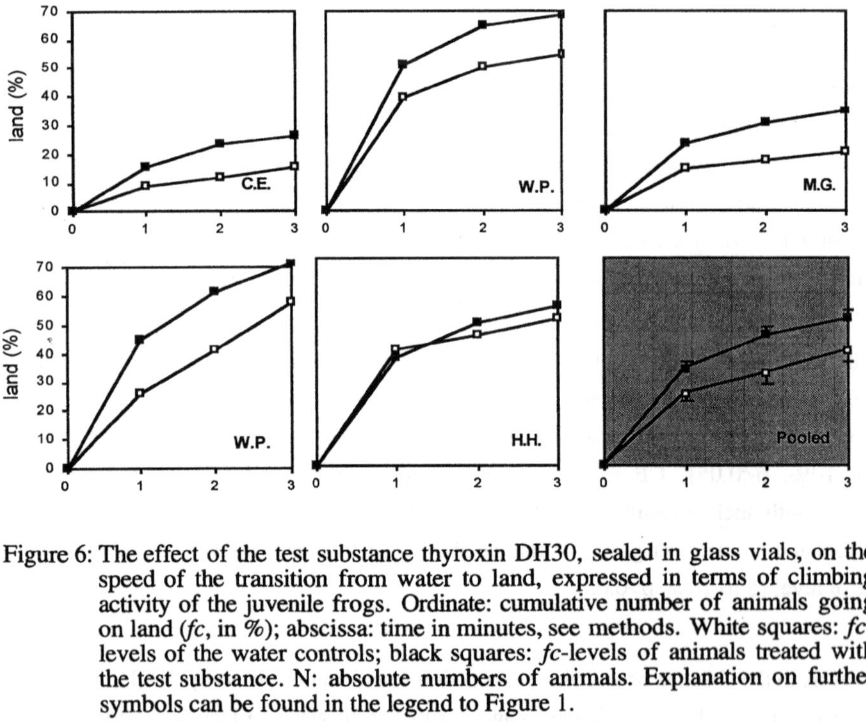

Figure 6: The effect of the test substance thyroxin DH30, sealed in glass vials, on the speed of the transition from water to land, expressed in terms of climbing activity of the juvenile frogs. Ordinate: cumulative number of animals going on land (*fc*, in %); abscissa: time in minutes, see methods. White squares: *fc*-levels of the water controls; black squares: *fc*-levels of animals treated with the test substance. N: absolute numbers of animals. Explanation on further symbols can be found in the legend to Figure 1.

Note: In taking account of the different number of animals per basin in the studies performed by the different researchers (e.g. 16, 18 or 20), the standard deviations given in the grey fields were always calculated sets of data where the number of animals per basin was the same (e.g. 18 animal). Thus, the standard deviations shown do not refer to the pooled data, but are representative of identically performed experiments.

In summary, long-term treatment with thyroxin DH30, sealed in glass vials leads to stimulation of metamorphosis to the land animal, observed in terms of climbing activity.

DISCUSSION

It is well known from experimental zoological physiology that conventionally produced thyroxin solutions in concentrations of 10^{-8} have a stimulating effect on amphibian metamorphosis. Explorative experiments indicate that this stimulation does not depend on whether the final concentration was reached via a single dose of an appropriately concentrated thyroxin solution at the start of the experiment, or via several doses added in intervals of 8 hours.

C.E	Graz 1 ll b		
	WDH30V	TDH30V	
	90	90	
0	0	0	-
1	8	14	-
2	11	21	*
3	14	24	-

W.P	Graz 2 ll b		
	WDH30V	TDH30V	
	26	26	
0	0	0	-
1	10	13	-
2	13	17	-
3	14	18	-

M.G.	Graz 1 ll r		
	WDH30V	TDH30V	
	72	72	
0	0	0	-
1	10	16	-
2	12	22	-
3	14	25	-

W.P	Graz 2 ll r		
	WDH30V	TDH30V	
	112	112	
0	0	0	-
1	29	50	*
2	46	67	*
3	64	80	-

H.H.	Vienna 1 hl r		
	WDH30V	TDH30V	
	80	90	
0	0	0	-
1	31	33	-
2	35	43	-
3	40	49	-

Pooled data			
	WDH30V	TDH30V	
	380	380	
0	0	0	-
1	89	123	*
2	118	167	**
3	147	189	**

Table 6.: DH30V, climb.

Table 6: Onset of the land stage in the experiments represented in Figure 6. For further information, see there. Note: this Table gives the means of the aggregate values of animals for five consecutive runs of the water-to-land experiment.

The solutions with final concentrations of 10^{-11} ff, which were tested in our study, in some, but not all experiments, deviated from current expectations with regard to the effects of thyroxin solutions. In cases of solutions prepared by gradual dilution *without* succussion, the concentration-effect curve normally aligned itself with that of the water control, as expected. However, by one researcher, with progressive dilution, a trend towards inhibition at the final concentration level of 10^{-13} ff has been reported [26]. In the case of solutions prepared *via gradual dilution and succussion*, in some, but not all experiments, inhibition already occurred at low levels of dilution (10^{-11}, DH6, and 10^{-13}, DH8).

These trends, however, may be in line with fundamental assumptions concerning hormesis [27] and homoeopathy. The concept of hormesis refers to unsuccussed dilutions especially of xenobiotic substances (substances biologically unknown to the organism) at the medium concentration level. It states that dilutions can cause effects which are opposite to those of the original solution from which they are derived [27]. These trends point to the possibility that the production process used with conventional hormonal preparations may neutralize, or even reverse, the latter's intended effect [6]. They are also noteworthy, and should be discussed, with regards to the methods of medicine production commonly used in homoeopathy.

Homeopathy's "potency rule" [7] states that such effect reversals may be intensified by preparing the dilution according to a standardized procedure of gradual dilution and strong succussion [7, 27].

The experiments involving thyroxin DH30 conducted in study 2 provided additions to the corresponding results of study 1. In particular, it seemed that the thyroxin DH6- and DH8-induced effect reversal observed in (1) intensifies at the extremely high dilution level (DH30). An inhibiting effect was most apparent with 48-hourly doses in autumn [13] (2.1.1), where natural development progresses relatively slowly. Development may also be stimulated when metamorphosis progresses relatively quickly, as a result of the season (summer). The fact that, in appropriate seasonal conditions, thyroxin DH30 can apparently also develop its inhibiting effect through glass walls, (2.1.2, [28]) and that it is possible to store the underlying information on a data carrier [29,44], confirms the suspected non-molecular nature of the effect of this solution, which was in principal prepared according to the homoeopathic pharmacopoeia [30].

Study 3 provided further confirmation of the results of study 2.2. By way of transference via an electronic amplifier, it seems to be possible to imprint information of the thyroxin solution DD3 onto a water sample. The result was, again, an inhibition of metamorphosis activity [44].

In all experimental variations looked at in studies 1, 2 and 3, to different degrees, the cumulative frequency of four-legged animals started to fall behind that of the control animals in the course of treatment when the respective data were pooled. However, it has to be stressed that two comparatively large experiments, one with DH6 and one with DH30, sealed in glass vials, could not confirm the respective overall results with DH6 and with sealed DH30, and that one experiment with DH8 was contradictory to the overall result with DH8. We have no obvious explanation for the apparent trend that results initially obtained for a certain experimental variant generally turned out more significant than in subsequent attempts at reduplication by other researchers or in other laboratories. With regard to this case and to similar findings in analogous cases one might hypothesize that the senior researcher on account of his experience may be using some sort of preconscious know-how to optimise the experimental protocol in details too small to be standardised [50]. Directly researcher-dependent effects have also been discussed in the literature [50,51].

Though we have no really incontestable findings to offer, the present studies in our view do nevertheless suggest that information of thyroxin may be transferred in several ways: it may pass over to the solution via gradual dilution in water and succussion according to a special procedure, it may exert an influence through the glass wall of a sealed glass vial, it may be transferred via electronic switching circuits and stored on CD, or it may be directly transferred electronically. This strengthens the assumption [16] that this bio-information is of an electromagnetic nature, or rather contains an electromagnetic aspect. The underlying order of such (long-range) fields, which are correlated as distinct from noise from other sources, should be described in terms of electrodynamics or quantum physics (see review article by Schulte this volume).

An explorative experiment showed that the simultaneous administration of the succussed solution DH8 reduces the stimulating effect of DD4. The finding indicating that metamorphosis activity in frogs, which has been additionally stimulated by exogenous thyroxin, can be slowed down by a dilution of that same substance, accords with homeopathy's so-called principle of similarity [7, 38, 39, 50]. In its simplest form, this rule states that a living thing, which has been poisoned by a certain substance, can be cured by the administration the same substance in crude or in highly diluted form [39]. The results suggest that the starting positions determined by a) exogenously administered synthetic L-thyroxin-sodiumpentahydrate, and b) the tadpoles' natural development, need to be differentiated.

With the continuation of treatment as in study 2.1.2 (constant presence of thyroxin DH30 in glass vials) beyond the juvenile stage, metamorphosis activity was accelerated rather than inhibited. In contrast, the respective preceding effect prevailed, at least in the autumn experiments, when thyroxin DH30 was continuously added in drop form (studies 1.2 and 2.1) until the animals reached the juvenile stage, i.e. the completion of metamorphosis was accelerated by 8-hourly doses and inhibited by 48-hourly doses. An inhibiting effect on climbing activity (water-land) was also observed when as yet untreated juveniles were treated with thyroxine DH30 on a short-term basis (minutes), irrespective of whether the test substance was added in drop form directly (statistically significant) , or locked into vials (trend). The results of study 5.1 can be discussed to the effect that, next to the animals' starting position, it is the frequency at which signals are given, which decides whether the information of thyroxin DH30 leads to inhibition or stimulation.

It seems reasonable to consider these results in relation to some typical characteristics of homoeopathic medicine. There are many reports in homoeopathic literature about high dilutions, which are capable of either intensifying or reversing the effects of their original substance in molecular concentration, depending on the experimental and clinical conditions. The amphibian model described here (high dilution level) appears to be especially suitable for demonstrating such a bivalent effect, since in this case the effect is created using one and the same experimental design.

In order to describe such effects, which are identical or opposite to those of the original substance, the terms "orthotaxic" and "antitaxic" are used (which were coined by P. Fisher in an unpublished statement).

Both of the possibilities discussed here can also be observed in humans, and have gained fundamental importance in the area of homoeopathic pharmaceutics.

Orthotaxic and antitaxic effects

Two unusual features of the effect of thyroxin DH30 have been described in the results section on results, and in [4]:

1. In experiments progressing relatively slowly, a) an initial further *inhibition* of the development towards the four-legged stage may be followed by b) a *stimulation* of the development towards the land-stage.

2. In experiments progressing relatively quickly, an initial and continued stimulation of development.

The fact that a substance in highly diluted form can have an effect of the same kind as the original substance in molecular concentration is not a new finding, but rather the basis of all homoeopathic drug testing of high dilutions [38, p.221; 40; 7]. [41] contains a report on a case of drug testing for thyroidinum, using healthy experimentees. This study was based on the hypothesis that a high dilution should, amongst other things, cause similar effects as a thyroxin solution of molecular concentration ("empirical homoeopathic poisoning study" or pathogenetic study). As in all homoeopathic drug testing, the study's aim was to identify the symptoms caused by this substance, in order to deduce, by way of the principle of similarity, those symptoms whose presence in a patient would indicate that treatment with thyroidin dilutions would be promising. The voluntary experimentees took the homoeopathically prepared thyroidin dilution DH30 in short intervals. It was a double blind experiment. Many volunteers in the verum group showed symptoms such as unrest and hyperactivity. This was attributed to the regular intake of thyroidine dilution.

If one wishes to get a general idea about a living system's defence reaction to the intake of a high dilution, or its original substance, an overall view of the concepts of the so-called rebound effect, of hormesis, and of homoeopathic fundamental research proves to be revealing [42]. A dynamic process, which manifests itself as a reversal of the primary toxic effect (recovery, see [39]), is the organism's reaction to the impact of a strong pharmacon or poison. High concentrations may here lead to effects which are opposite to those brought about by low concentrations (Arndt-Schultz law, [27]).

"Recovery is the organism's reaction to aggression. It uses specific, suitable means in order to fight the aggression. Such a reaction may also be caused by an appropriate pretreatment, or by treatment after poisoning. The pretreatment triggers a learning process in the organism, while in the after-treatment, the recovery process is accelerated via a dose of additional information. The reaction may be caused by the molecule itself, or directly by the information contained in a high dilution of the toxic molecule. Both trigger a learning process" [42].

Attempts have also been made to detect in our study a parallel to various detoxification studies (F. Wiegant, [13]). Generally, during detoxification experiments, an organism is initially poisoned by a high, toxic dose of a substance, and subsequently treated with a low dose, or a highly diluted, succussed solution, of the same substance. In a multitude of studies involving various organisms, the detoxification process could thus be accelerated in comparison to the respective control groups [see the article Classification of Fundamental Research into Homoeopathy].

In the course of our research there was only one case of hyperstimulation ("poisoning") of the animals with thyroxin.

It seems obvious to assume that the orthotaxic and antitaxic effects achieved with the homoeopathically prepared high thyroxin dilution are to be attributed to the same laws which are the basis of the orthotaxic and antitaxic effects observed in homoeopathy for the past 200 years.

Resonant frequencies of the test solutions

In laboratory collaboration with C.W. Smith, samples of the trial substances used in this study were examined as to their possible natural resonant frequency patterns. Resonance-related phenomena had already been observed in a preceding study [16, p.203; 42].

By scanning several oscillators, a frequency range of 0,1 Hz to 10 MHz was examined, in the course of which it was observed that certain frequencies seemed to lead to an interaction with the trial substance introduced into the field. This coupling of resonances was monitored via a standardized biological reaction (test person's microtremor).

In the case of gradually diluted and succussed trial substances, each of the successive preparatory steps lead to the appearance of two additional (higher) natural frequencies [16, pp. 204].

Table 7 (overleaf), left, shows the natural resonant frequency levels obtained in seven independently conducted experiments (frequency range 1-100 Hz, corresponding to thyroxin DH12).

Table 7, right, shows the levels of the trial substances 2.2 (intermediate storage on a data carrier) and 3 (transference via bio-resonance device) (frequency range 1-100 Hz, one sample point each). In this range, untreated water only shows a resonance at

8,0 Hz. In cases where a partial formation of clusters of levels occurs in samples 1-7 (for example around 3, 7, 10, 12, 16, 19, 90 Hz), corresponding frequencies can usually also be found in the electronically processed trial substances.

Keeping in mind the limitations mentioned in [16, p.206], these findings seem to be helpful in facilitating both an expanded theoretical concept of information transference, and thus also the possibility of goal-orientated physiological experiments. It is hoped that the automatic (technical) determination of such natural frequencies, as well as biological experiments involving these frequencies, will provide essential information (see also [16, p.215-218]).

Frequencies found by C.W. Smith have been generously arranged and transferred onto CD by Dr. Robert Höldrich at the Institute of Electronic Music in the College of Music, Graz.

As could be shown by way of experiments, phenomena of life are usually linked to the emission of electromagnetic waves. These emissions are fundamentally connected to all biological phenomena. Living systems possess highly developed electromagnetic systems of bio-communication. The correlation or coherency of these signals enables extraordinary characteristics of living systems to develop, e.g. the maximum possible sensibility towards low energy information at a maximum possible signal/noise ratio [16, p.250].

Specific information may be transmitted via electromagnetic frequencies when they are in phase, which distinguishes them from a multitude of unordered influences, and renders them similar to a technical laser [16, Outlook].

During the laboratory collaboration between C.E. and C.S. it was possible, through the influence of various frequencies (from Table 10), to clearly stimulate or slow down spontaneous, rhythmically swelling and fading emission activity in amphibian larvae in the low frequency range (1-100 Hz) (currently being prepared for publication). It seems reasonable to assume that this emission activity is typical for all organisms [42; measuring method as in 16, pp. 203]. The animals may be introduced either into an

-------	*successed dilutions*			-------------------------			*via*	*via*
1	*2*	*3*	*4*	*5*	*6*	*7*	*CD*	*amplifier*
1,3	1,5	1,5	-	-	-	-		
2,6	3,0	-	-	-	3,0	-	3,0	3,0
-	-	4,0	4,0	-	-	-		
4,8	5,0	-	-	-	-	-		
6,3	6,5	-	-	7,0	7,0	7,0	7,0	7,0
8,4	8,5	8,0	-	-	-	-		
-	-	-	9,0	-	-	-		
10,0	10,0	-	-	10,0	-	-		
-	12,0	12,0	12,0	-	-	-	12,5	13,5
-	16,0	-	15,5	14,5	16,5	16,0	16,5	
18,0	-	19,0	19,0	18,9	19,0	19,0	19,0	18,0
-	-	-	-	22,0	-	-		
42,0	45,0	-	-	-	-	-	47,0	45,0
-	-	-	-	55,0	50,0	50,0		
67,0	70,0	60,0	65,0	-	-	-		
89,0	90,0	-	-	-	90,0	90,0	87,0	90,0

Table 7: Resonant frequencies (Hz) of a trial substance from studies 2.1, 2.2 and 3. Explanations can be found in the text.

Oscillations of the living system and influence of trial substance-typical frequencies

appropriate electrical field, or into a vector-potential field (16, p.193), which alters the speed of the emissions' swelling and fading by up to the factor 10. The animals' emission activity is synchronized as is shown in the following protocol.

Minute 0: vector-potential field, generated via oscillator and toroid coil, is switched on. Minute 10: synchronization has occurred. Three biological frequency emissions take place before minute 40. Minute 40: vector-potential field is switched off. Up to minute

80: emission activity does not materialize, beginnings of desynchronization. Minute 82: vector-potential field is switched on, followed by a frequency emission by the animals. The natural incidence of emission activity without interference lies outside the time frame of observation. In order to determine the biological frequencies emitted, the oscillator was switched on intermittently between Minutes 40 and 82 for short intervals.

This synchronization may be neutralized by way of optical separation. Further experiments showed that the animals needed to have optical contact with each other for wavelengths of up to 495 nm (yellow light, 6×10^{14} Hz), in order for their biological emissions to synchronize in the low frequency range. This finding could point to secondary communication via biophotons [see Bischof's article in ref. 49].

It seems that an interpretation of the findings discussed here is most feasible when the biological system is viewed as a macroscopic system (Ho, this volume). In this context, we would also like to point out the possibility of viewing the substance 'water' as a macroscopic quantum system.

ACKNOWLEDGMENTS

Independent experiments were carried out by Christian Endler and Waltraud Pongratz as well as: Jörg Alex, Faculty of Medicine at the University of Tübingen; Daniela Dieterle, Faculty of Medicine at the University of Tübingen; Michael Gehrer, Ludwig Boltzmann Institute of Homoeopathy, Graz; Helge Hilgers, Zoological Institute at the University of Vienna; Egon Lauppert, Research Unit for Low Energy Bio-information, Graz; Christa Lukitsch, Zoological Institute of the University of Vienna; Christina Vinattieri, Faculty of Holistic Medicine, University of Urbino & IDRAS, Turin; Karl Waltl, Boltzmann Institute of Homoeopathy, Graz; Fred Wiegant, Department of Molecular Cell Biology, University of Utrecht, Netherlands; Roland van Wijk, Department of Molecular Cell Biology, University of Utrecht. For a short period of time, Conrad Heckmann conducted some studies at the University of Graz's Zoological Institute for the purpose of preparing his biology thesis for the Faculty of Medicine at the University of Tübingen. Andrea Nograsek from the Office for Environmental Protection under the State Government of Styria carried out the coding for control experiments of type 2.1.

Documentation of the results obtained by the individual independent participants in the experiments was undertaken by Roland Brandmeier from the Biometric Center for Therapy Studies, Munich, FRG.

Our special thanks goes to Thomas Kenner from the Physiological Institute at the University of Graz, Austria, to Max Haidvogel from the Boltzmann Institute of Homoeopathy, Graz, as well as to all those whose help made our experiments possible.

References

1. Giersberg H. Hormone. Springer, Berlin 1972.
2. Ganong W.F. Lehrbuch der Medizinischen Physiologie. Springer, Berlin 1974.
3. Wehner R., Gehring G. Zoologie. Thieme, Stuttgart 1990
4. Endler P.C., Pongratz W., van Wijk R., Wiegand F.A., Waltl K., Gehrer M., Hilgers H. A zoological example on ultra high dilution research. In 16.
5. Pitt-Rivers R., Trotter W.R. The Thyroid Gland. Butterworth, London 1964.
6. Endler P.C. Wirkungsumkehr des Hormons Thyroxin. Zur Frage ungewollter Wirkungen von Pharmazeutika. Jubiläumsfonds der Österreichischen Nationalbank, Projekt Nr. 5402. Durchgeführt am zoologischen Institut der Universität Graz.
7. Dellmour F. Homöopathie. Facultas Universitätsverlag, Wien 1991.
8. Endler P.C., Pongratz W., Kastberger G., Wiegant F.A.C., Schulte J. The effect of highly diluted agitated thyroxine on the climbing activity of frogs. Vet. Hum. Tox. 1994;36,1:56-59.
9. Bastide M., Daurat V., Doucet-Jaboeuf M., Pélegrin A., Dorfman P. Immunomodulator activity of very low doses of thymulin in mice. Int. J. Immunotherapy 1987;3:191-200.
10. Hadji L., Arnoux B., Benveniste J. Effect of dilute histamine on coronary flow of guinea pig isolated heart. FASEB J. 1991;5:A1583.
11. Benveniste J., Arnoux B., Hadji L. Highly dilute antigen increases coronary flow of isolated heart from immunized guineapigs. FASEB J. 1992;6:A1610.
12. Youbicier-Simo B.J., Boudard F., Meckaouche M., Bastide M., Baylé J.D. The effects of embryonic bursectomy and in ovo administration of highly diluted bursin on adrenocorticotropic and immune response of chicken. Int. J. Immunotherapy 1993:IX:169-190.
13. Endler P.C., Pongratz W., Smith C.W., Schulte J. Nonmolecular information transfer from thyroxine to frogs with regard to homoeopathic toxicology. Vet. Hum. Tox. 1995;37/3:259.
14. Pongratz W., Endler P.C., Poitevin B., Kartnig T. Effect of extremely diluted plant hormone on cell culture. Proc. AAAS Ann. Meeting 1995.
15. Pongratz W., Endler P.C., Nograsek A. Effect of extremely diluted trace substance on wheat growth. Proc. AAAS Ann. Meeting 1996.
16. Endler P.C., Schulte J. (eds.). Ultra High Dilution. Physiology and Physics. Kluwer Academic Publishers, Dordrecht 1994 (ISBN 0-7923-2676-8).
17. Gosner K.L. A simplified Table for staging anuran embryos and larvae with notes on identification. Herpetologica 1960;16:183-195.
18. Endler P.C., Pongratz W., van Wijk R., Kastberger G., Haidvogl M. Effects of highly diluted sucussed thyroxine on metamorphosis of highland frogs. Berlin J Res Hom 1991;1:151-160.
19. Endler P.C., Pongratz W., Smith C.W., Schulte J., Senekowitsch F., Citro M. Non-molecular information transfer from thyroxine to frogs. By means of homoeopathic preparation and electronic processing. In: Bastide M. (ed.). Signals and Images. Kluwer Academic Publishers, Dordrecht 1996.
20. Citro M. Vom Pharmakon zur Frequenz; elektronischer Medikamententransfer. Proc. II int. Symp. Biokybernetische Medizin, Würzburg 1991.
21. Citro M. TFF dal farmaco alla frequenza. Vivibios 1992;II,3:25-30.
22. Citro M., Pongratz W., Endler P.C. Transmission of hormone signal by electronic circuitry. Poster pres. AAAS Ann. Meeting 1993.
23. Aissa J., Litime M.H., Attis E., Benveniste J. Molecular signalling at high dilution or by means of electronic circuitry. J. Immunolog. 1993;150:A146.
24. Benveniste J., Aissa J., Litime M.H., Tsaegaca G.T., Thomas Y. Transfer of the molecular signal by electronic amplification. FASEB J. 1994;8:A398.

25. Citro M., Smith C.W., Scott-Morley A., Pongratz W., Endler P.C. Transfer of information from molecules by means of electronic amplification. In 16.
26. Endler P.C., Lauppert E., Heckmann C. Inversion of hormone effect by agitation. FASEB J. 1995:A, angenommen.

Lauppert E., Endler P.C. Enhanced inversion effect of thyroxine 10^{-10} $^{-13}$ by agitation. Curative effect following hyperstimulation in frogs. In: Taddei Ferretti C. (Hrsg.). High Dilution Effects on Cells and Integrated Systems. World Scientific, Singapore 1996.
27. Oberbaum M., Cambar J. Hormesis: dose-dependent reverse effects of low and very low doses. In 16.
28. Endler P.C., Pongratz W., van Wijk R., Waltl K., Hilgers H., Brandmaier R. Transmission of hormone information by nonmolecular means. FASEB J. 1994;8:A400.
29. Senekowitsch F., Endler P.C., Pongratz W., Smith C.W. Hormone effects by CD record /replay. FASEB J. 1995;9:A12161.
30. Homöopathisches Arzneibuch. Deutscher Apothekerverlag, Stuttgart und Govi Verlag, Frankfurt 1991.
31. Citro M., Endler P.C., Pongratz W., Vinattieri C., Smith C.W., Schulte J. Hormone effects by electronic transmission. FASEB J. 1995;9:A12025.
32. Endler P.C., Citro M., Pongratz W., Smith C.W., Vinattieri C., Senekowitsch F., Übertragung von Molekül-Information mittels Bioresonanz-Gerät im Amphibienversuch. Erfahrungsheilkunde 1995;3:186-192.
33. del Giudice E., Preparata G., Vitiello G. Water as a free electric dipole laser. Phys. Rev. Lett. 1988;61:1085-1088.
34. del Giudice E. Is the 'memory of water' a physical impossibility? In 16.
35. Schulte J. Conservation of structure in aqueous ultra high dilutions. In 16.
36 Berezin A.A. Ultra high dilution effect and isotopic self-organisation. In: Ultra High Dilution, Endler P.C., Schulte J. (eds.). Kluwer Academic Publishers, Dordrecht 1994:137-169.
37. Schulte J., Endler P.C. Preliminary elements of a theory on ultra high dilution. In 16.
38. Andersch P., Endler P.C. Glossary on homoeopathy. In 16.
39. van Wijk R., Wiegant F.A.C. Cultured mammalian cells in homeopathy research: the similia principle in self-recovery. University Utrecht, Biological Faculty, Utrecht 1994.
40. Walach H. Scientific proving of an ultra high dilution. In 16.
41. Panos M., Rogers R., Stephenson J. Thyroidinum, A Proving (Hyganthropharmacology). J. Americ. Inst. Hom. 1964; July-August:197-207.
42. Bastide M. In: Taddei-Ferretti C. (eds.). High Dilution Effects on Cells and Integrated Systems. World Scientific, Singapore 1996.
43. Smith C.W. Measurement of the electromagnetic field generated by biological systems. Neural Network World 1995;5:819-829.
44. Senekowitsch F., Citro C., Vinattieri C., Pongratz W. Smith C.W., Endler P.C. Amphibien-metamorphose und die elektronische Ubertragung von Bioinformation. In: Stacher, A. (Hrsg.) Niederenergetische Bioinformation. Facultas Universitätsverlag, Wien, im Druck.
45. Heckmann C. Einfluß homöopathisch und nicht homöopathisch hergestellter Thyroxinlösungen auf die Metamorphosegeschwindigkeit beim *Grasfrosch (Rana temporaria* L.). Thesis, Faculty of Biology, Tubingen University, 1997.
46. Alex J. Doctoral thesis, Faculty of Medicine, Tubingen University, in preparation.
47. Lukitsch C. Thesis, Faculty of Philosophy, Vienna University, in preparation.
48. Dieterle D. Doctoral Thesis, Faculty of Medicine, Tubingen University, in preparation.
49. Endler PC and Schulte J, Homöopathie - Bioresonanztherapie, Verlag Maudrich 1996 (ISBN 3-85175-6681).

FUNDAMENTAL RESEARCH INTO HIGH DILUTION EFFECTS.

A CLASSIFICATION OF NON-CLINICAL RESEARCH TOPICS

Christian Endler

Basic tenets of fundamental research on homoeopathy

Observation of physiological effects; 'similarity' and 'potency' rule

A large variety of studies with regard to fundamental research on homoeopathy has been presented up to now. The studies on the interaction of an organism and a test substance can in principle be divided into two complementary parts.

Studies were performed *on the reactional cybernetics of living systems* with application of either conventionally prepared (diluted) test substances or test substances prepared according to a 'homoeopathic' protocol of step by step dilution and agitation ('potencies').

This type of studies is related to the *similarity principle* which all clinical homoeopathy is based on. The similarity principle states that a substance which may cause symptoms in a healthy organism, may be applied to a diseased or ill organism in a curative sense if this organism shows identical symptoms (symptoms similar to the former). The stimulation of an irritated autoregulative recovery by application of the *similarity principle* is considered to be the essence of homoeopathy. In a basic (reduced) form, this rule concerns the curative effect of an agent at high dilution after intoxication of a living system with the same agent at low dilution - 'isopathic principle' in a modern sense, van Wijk and Wiegant, 1994 -).

Research on homoeopathy, however, may focus more *on the test substance applied* (*potency* research). Biological studies were either performed in order to demonstrate the difference between 'conventionally' diluted substances and homoeopathic potencies of the same substances, mostly at the level of low dilutions, or in order to describe biological effects caused by (often extremely) high potencies tested versus neutral control.

J. Schulte and P.C. Endler (eds.),
Fundamental Research in Ultra High Dilutionand Homoeopathy, 189-213.

The traditional *potency principle* assumes that, provided the similarity rule is met, a homoeopathic drug is more active ('sharp'), the more the initial mother substance is diluted in the standardised process of stepwise dilution (e.g. 1:10 or 1:100, respectively) and agitation (intermediate succussion). Furthermore, typical effects (e.g. stimulation versus inhibition of a physiological process) seem to be related to typical ranges of dilution.

More studies on a few reliable models linking the gap between the areas of low and high dilution are needed in order to have more knowledge about the interconnection of the *similarity principle* and the *potency principle.*

Biophysical explanation

Further studies concern *the biophysical interaction* of an organism and a test substance.

Physics approach

The *physical properties* of the potencies themselves are subject of further studies.

Common aims of fundamental research

It is obvious that a synopsis of all research results should ideally point towards a consistent description of the reaction cybernetics of the living system *and* of the changes occurring during the dilution process of the test substance.

In the following, existing studies are classified with regard to the stimulus or stimuli applied: this may be either only one single stimulus or two consecutive stimuli. In the latter case, the respective mother tinctures may be identical or different, if a diluted substance is used. Furthermore, the dilutions may be prepared in different ways (Table 1).

This differentiated way of presentation is chosen in order to show that homoeopathy is a genuine type of science in itself, with its own research strategies. Furthermore, it was chosen in order to stimulate more research along these lines, perhaps leading to a wider understanding of the laws of nature in general.

In this text, the symbol '##' indicates that a study met the criteria for publication in conventional (indexed) scientific journals, i.e. that the respective paper underwent a peer-review process. This classification may exclude important work that, for different reasons, was not submitted to conventional media. The symbol '#' indicates that a study met the criteria for discussion in conventional journals, i.e. that the content of the study was explicitly discussed in such media or that a relevant abstract was printed in a peer-reviewed journal.

Specially, some older studies may have to be added. On the other hand, some selected clinical studies were entered into the systematisation for didactical purposes.

It has to be pointed out that this list of studies is not exhaustive. Furthermore, the paper remains incomplete with regard to research on the preparation process of homoeopathic potencies, quality control and storage conditions.

	classification of study	*source of 1st and 2nd stimulus:*			*type of observation*
		1st	*2nd*		
A	conv. pharmaceutics			**L**	- typical effect
B	hormesis		**H**	-	inversion of effect
C	tolerance	**i**	**L**	**L**	protection
I	isopathy		**P**	-	exploration of effects
II	"	**i**	**L**	**H**	cure
III	"	**i**	**L**	**P**	detoxification
IV	"	**i**	**P**	**L**	protection
V		**i**	**P**	**P, P**	provocation of symptom
				(hom. drug proving)	
VI	homoeopathy	**d**	**L**	**L**	protection
VII	"	**d**	**L**	**H**	cure
VIII	"	**d**	**S**	**L**	a) symptom
				L	b) cure
IX	"	**d**	**L**	**L**	a) symptom
				H	b) cure
X	"	**d**	**L**	**L**	a) symptom
				P	b) cure
XI	"	**d**	**S**	**L**	a) symptom
				H	b) cure
XII	"	**d**	**S**	**P, P**	a) symptom
				P	b) cure

Table 1: For legend, see next page

Legend to Table 1: Effects described after application of one stimulus or two consecutively applied stimuli on a living system.

A,B,C; I-XII, different types of studies. For classification, see below.

Source of 1st and 2nd stimulus: in case a dilution is used, the respective mother tinctures may be i, identical, or d, different.

Type of stimulus: the 1st and /or the 2nd stimulus may be at low dilution (crude substance or mother tincture, L); or at high dilution (any degree of dilution of the mother tincture, H); or at high dilution in potentised form (P); or the initial state may be a stressed or pathological state (illness) (S). Theoretically, 24 combinations (types of studies) are possible.

The different types of studies found in literature are:

A **conventional pharmaceutical research**

occurrance of a typical effect of an agent as known from conventional pharmacology

B **research into hormesis**

inversion of this effect at high dilution of this agent, known as 'hormesis'

C **research into self-tolerance**

protective effect of a 1st stimulus with regard to a consecutive 2nd application of the same noxious agent

fundamental research on isopathy including the empirical homoeopathic intoxication design in drug proving

I different effects of potentised agents that remain to be further explored

II stimulation of recovery of the living system by the 2nd stimulus

III a detoxification by the 2nd stimulus

IV a protective effect of the 1st stimulus

V the provocation of substance specific (illness) symptoms by frequent repetition of one and the same stimulus (empirical homoeopathic intoxication design for drug proving)

fundamental research on homoeopathy and therapeutic application of homoeopathy

VI a protective effect of the 1st stimulus, even if this stimulus is of a different nature from the 2nd stimulus

VII a curative effect of the 2nd stimulus, even if this stimulus is of a different nature from the 1st stimulus

VIII the initial finding of S. Hahnemann on 'homoeo'-pathy:

(α) in two different individuals 1 and 2, similar typical symptoms may occur due to a pathological process (stimulus 1) or due to an intoxication (stimulus 2), respectively

(β) stimulus 2, e.g. a crude substance, exerts a curative effect on the individual 1

IX recent cell biological finding. As in (VIII), with stimulus 2 applied at a high dilution

X immunological finding. As in (VIII), with stimulus 2 applied at a high dilution in potentised form

XI examples from modern conventional medicine on the use of diluted agents

XII the frame of homoeopathic medicine, based on the similarity between stimulus 1 and 2 (simile principle) and on the use of the agent (stimulus 2) at high dilution in potentised form (potency principle). P, P means the previous provocation of symptoms in a drug proving (see V).

For further details, see text.

A Conventional pharmaceutical research

Conventional pharmaceutical research describes typical physiological effects of agents at relatively low dilutions. Normally, these effects are either stimulatory or depressive.

B Research on hormesis

The Arndt-Schulz principle states that weak stimuli stimulate vital activity, medium stimuli may support this activity, while the strongest stimuli may annul vitality (Martius 1923). Put simply, hormesis means that in many cases a stimulatory effect occurs in living systems after exposure to a high dilution of an otherwise toxic agent, whereby stimulation may even mean an improvement of health or an enhancement of longevity (Oberbaum and Cambar 1994). In various fields of biology, such dose-dependent reverse effects (biphasic dose /response relations) have been described and a large number of papers has been presented in conventional media (for survey see Linde 1991; Popp 1994; Oberbaum and Cambar 1994).

Such reversed biological effects in various ranges of concentration of the same agent represent the basic principles of the *similarity* and the *potency principle* (Popp 1994). There is evidence that these effects become more marked on special preparation (homoeopathic potentisation) of the test substance (Larue et al. 1986 ##; Cal et al. 1988 ##; Lauppert and Endler 1996 - critically -). The application of the hormesis concept to homoeopathy is critically discussed due to biochemical findings (Dittmann et al. 1994a,b ##, ##).

C Research on tolerance

These studies concern the protective effect of an agent at low dilution with regard to a consecutive second application of the same agent at the same concentration.

In the so-called 'self-tolerance' studies, the same stressor conditions are applied twice in the same strength with a time interval in between. In various fields of biology, protective effects of the 1st stimulus have been described and a large number of papers has been presented in conventional media (see van Wijk and Wiegant 1994).

Aspects specially relevant for homoeopathy are, firstly, that one of the stimuli may be applied at high dilution (see II, III) and, secondly, the 1st and the 2nd stimulus may be of a different nature (see VII, 'cross-tolerance').

Research into isopathy

I Different effects of agents at high dilution in potentised form

A large number of studies has been performed in order to test the assumption that agents can exert biological activity even if applied at high and extremely high dilution, often beyond Avogadro's limit of theoretical O-molarity. The structure of these experiments is simple in comparison to the other models, since only one stimulus is applied. In all cases, the agent underwent a potentisation process.

In vitro and ex vivo studies

Boyd 1954 studied the influence of a mercury potency on the starch splitting activity of malt-diastasis. In this meticulous study performed with a very large number of samples over a period of several years, a stimulation of this reaction was found (#).

Harisch and Kretschmer 1988a,b studied the influence of potentised zinc on peritoneal cells of rats. An affection of histamine release was found (##).

Bonavida 1992, Bonavida et al. 1993 studied the influence of synergistic compounds on tumour cell growth. Typical effects were found (##).

Doutremepuich et al. 1987, 1990, 1994 studied the influence of potentised acetylsalicylic acid on blood platelet aggregation. In this type of study that was repeated over several years, a reduction of bleeding time was found (##).

Belon 1987; Boiron and Belon 1982; Poitevin et al. 1986, 1988; Sainte Laudy 1989; Sainte Laudy et al. 1986; Sainte Laudy and Belon 1993 studied the influence of potentised histamine or bee venom on the degranulation of basophils. Donors were either allergic persons or donor blood was sensitised with anti IgE antibodies (see also VIII). In these mutually confirmative studies, an inhibition of basophil degranulation was found (##).

Davenas et al. 1987 studied the influence of potentised quartz, a traditional homoeopathic remedy, on mice macrophages. A stimulation of their activity was found (##).

Davenas et al. 1988 (positive); Benveniste et al. 1991 (positive); Ovelgonne et al. 1992 (negative); Litime et al. 1993 (positive); Hirst et al. 1993 (negative) studied the influence of potentised anti IgE antibodies on the degranulation of basophils. In some, but not in all of these studies, a stimulation of basophil degranulation was found (##).

Weismann et al. 1992 a,b studied the influence of potentised antigen on IgM and IgG antibodies. A modulation of the immune response to KLH was found (see also VIII).

Petit et al. 1989 studies the influence of potentised substances on subcellular enzymatic activity. No effects were found (##).

Sukul and Zaghlool 1990 studied the influence of traditional homoeopathic remedies on the isolated ileum of rats. Hadji et al. 1991; Benveniste et al. 1992 studied the influence of different hormones on the activity of isolated perfused hearts. In these studies, an enhancing effect on the organs was found (#).

In vivo studies or zoological studies

Bastide et al. 1985, 1987; Doucet-Jaboeuf et al. 1982, 1984, 1985 a,b studied the influence of potentised thymulin on humoral and cellular immune response in (in some studies thymectomized) mice. Daurat et al. 1988 a,b studied the influence of potentised interferon on the immune response in mice. Youbicier-Simo et al. 1993 studied the influence of potentised bursin on (bursectomized) chickens. In all three types of the studies, a stimulation of the immune response was found (##).

König 1927 studied the influence of potentised metal salts on amphibian metamorphosis. Endler et al. 1994 a-d, 1995 a,b studied the influence of potentised thyroxine on amphibian metamorphosis and activity. In these mutually confirmative studies, stimulation and inhibition peaks were found at different degrees of dilution. The studies on the influence of thyroxine (see also III) were multicentrically repeated (##).

Sukul et al. 1986 studied the influence of traditional homoeopathic remedies on experimentally induced catalepsy in mice. A prolongation of the cataleptogenic effect was found (##).

Botanical studies

Kolisko 1926; Pelikan and Unger 1971; Jones and Jenkins 1981, 1983; Netien and Graviou 1978; Dutta 1989; Steffen 1984; Lehner et al. 1991; Pongratz and Endler 1994; Betti et al. 1994, Lauppert 1995 (##) studied the influence of potentised metal salts and other compounds on the growth of wheat and yeast, respectively. In these mutually confirmative studies, stimulation and inhibition peaks of growth were found at different degrees of dilution. Popp found effects at moderate, but not at high degrees of dilution. Two studies (Specciani 1988, negativ; Lauppert 1995, positive) were published as thesis. A survey is given by Dorfman 1993.

Boiron & Zervouacki 1962; Boiron & Marin 1965, 1971; Auquière et al. 1981, 1982; Pongratz 1995; Netien 1962, 1978; Netien et al. 1972; Noiret & Glaude 1979.

II Curative effect of an agent at high dilution after intoxication with the same agent at low dilution

In vitro studies

In this type of studies, a stressor condition (heavy metal, heat shock, oxidative stress) is applied to cell cultures in order to induce cellular recovery processes. Consecutively, it is determined whether after exposure to a damaging condition, cells can be stimulated by a high dilution of the same stressor, which would be ineffective in control conditions.

The aim of these experiments on cellular recovery processes is to analyze the stimulatory effect of low doses (= relatively high dilutions) of compounds which are applied according to the similia-principle in its most elementary form (isopathic principle, van Wijk and Wiegant 1994).

Van Wijk and Wiegant 1994; Ovelgönne et al. 1995 studied the influence of various stressors on cell cultures. In these mutually confirmative studies, a stimulation of autoregulative recovery was found (##).

III Curative effect of an agent at high dilution in potentised form after intoxication with the same agent at low dilution

In these studies as in (II), the similia-principle in its basic form is tested. Different from (II), the agent used for stimulation of recovery underwent a potentisation process. The reason for this is that it is assumed that the curative effect only occurs if or is more marked if the agent is potentised (potency principle).

In vivo or zoological studies

In this type of study, animals are detoxicated after experimental intoxication. Usually enhanced excretion of the toxin in the urine and stools is described. A meta-analysis is given by Roth 1991.

Cazin 1986; Boiron and Cier 1962 a,b, 1971; Boiron et al. 1968, 1978; Cazin and Gaborit 1983; Cazin et al. 1987; Chaoui 1988; Gaborit 1987 studied the influence of potentised arsenite after arsenite intoxication of rats. Mouriquand et al. 1961, 1962, 1975 used pigeons and Lapp et al. 1955, Lapp and Wurmser 1958; Wurmser 1984 used guinea pigs. In these mutually comparable studies, enhanced detoxification was described (##).

Fisher and Capel 1982 (positive); Fisher et al. 1987 (negative, - ## -) studied the influence of potentised lead on lead-intoxicated rats.

Herkovits et al. 1989, 1990 studied the influence of potentised cadmium on cadmium-intoxicated frog spawn. Decrease of letality was found (##).

Endler et al. (1994b, 1995b, see I) discussed their multicenter results on the inhibition of metamorphosis by potentised thyroxine with regard to the naturally enhanced thyroxine level of the amphibia. Lauppert and Endler 1996 studied the influence of potentised thyroxine on thyroxine hyperstimulized amphibia. In this study, a curative effect was found.

Botanical studies

Netien et al. 1965; Boiron & Marin 1967; Noiret & Glaude 1976; Projetti et al. 1985 studied the influence of potentised copper salt of plants previously intoxicated with copper salt . A curative effect was found.

Human studies

Paterson 1944 studied the influence of potentised mustard gas on mustard gas burns of the skin. Therapeutic (as well as prophylactic) effects were found.

Reilly et al. 1986 studied the influence of potentised grass pollen on hypersensitive persons exposed to those allergens. A therapeutic effect was found (##).

Gardes 1989 studied the influence of the potentised chemotherapeutic substance nalidixine acid on persons treated with nalidixine acid. Enhanced excretion was found. The study was published as a thesis.

IV Protective effect of an agent at high dilution in potentised form with regard to a consecutive second application of the same agent at low dilution

In these studies as in (III), the similarity principle in its basic form as well as the potency principle are investigated. In contrast to (III), the potentised agent is applied before the noxious agent.

In vitro study

Delbancut et al. 1993 studied the effect of potentised cadmium on cadmium-intoxicated renal tubular cell cultures. A protective effect was found.

Zoological studies

Cambar 1983 a,b; Guillemain et al. 1984; Larue and Cal 1985, Larue et al. 1986; Cal et al. 1986, 1988 a,b; Larue et al. 1985 studied the influence of potentised mercury on mercury-intoxicated mice. In these mutually comparable studies, a protective effect was found.

Souza Magro et al. 1986 studied the influence of the potentised antibiotic gentamycin on gentamycin-treated rats. A protective effect on nephrotoxicity was found.

Labonia et al. 1986 studied the influence of a potentised snake poison on mice intoxicated by this poison. A protective effect was found.

Botanical studies

Projetti et al. 1985 studied the influence of potentised copper salt on copper salt intoxicated lentils. A protective effect was found.

V Provocation of substance specific (illness) symptoms by frequent repetition of the application of one and the same agent at high dilution in potentised form

In these studies, again as above, one and the same agent is applied at different times. However, all stimuli (usually much more than two in this type of study, and usually diluted beyond Avogadro's value) underwent a potentisation process. Interestingly, while the effect of one single application of the potency was curative or protective, i.e. opposite to that of the agent at low dilution in (III) and (IV), in this type of study, frequent application led to disease or illness symptoms or an aggravation of symptoms.

Zoological study

Endler et al. 1994a studied the influence of potentised thyroxine on amphibia. When the extremely diluted agent was applied at short intervals, a stimulation of thyroxine-controlled metamorphosis and hyperactivity was found (see also I).

Human studies

are described in detail in contributions to drug provings (empirical homoeopathic intoxication studies). The intoxication symptoms are the basis of homoeopathic manuals.

Research into homoeopathy

In these studies, again as above, two stimuli are applied. In contrast to the protocols concerning the similarity principle in its basic form (i.e. the isopathic principle, see II-IV), here, the 1st and 2nd stimulus are of a different nature. The studies thus concern the similarity principle in its complex form.

VI Protective effect of an agent at low dilution with regard to a consecutive application of a different agent at low dilution

Different to the so-called 'self-tolerance' studies (C), in these 'cross-tolerance' studies, the protective effect is exerted by a 1st stimulus that is of a different nature than the 2nd stimulus. A very large number of conventional articles is present on this topic (see van Wijk and Wiegant 1994).

In vitro study

These studies are especially relevant to homoeopathy when an observed stressor-specificity is shown to result in a stressor-specific development of tolerance, such as was outlined by van Wijk and Wiegant 1994.

VII Curative effect of an agent at high dilution in potentised form after intoxication with a different agent at low dilution

These studies are closely related to the detoxification protocols (III) on the similarity and potency principle. However, the curative agent is not a potency of the noxious agent, but one of an entirely different agent.

Zoological study

Aubin et al. 1979, 1980; Bildet et al. 1975, 1977, 1978, 1984 a,b,c; Andresen 1985; Harisch and Kretschmer 1988 a,b, 1990; Harisch et al. 1984, 1986, 1987, 1990; Sur et al. 1990 studied the influence of traditional homoeopathic remedies (e.g. phosphorus) on CCl_4 induced liver cell damages in rats. A curative effect is reported (Harisch et al., some ##). One study (1985) was published as a thesis.

Santini et al. 1991 studied the influence of potentised copper on neostigmine induced intestinal hypertransit. A reduction of the neostigmine action was found (copper administered alone showed no effect) (##).

Veterinary study

Baumans et al. 1987 studied the influence of a traditional homoeopathic remedy on induced hypercholesterolaemia in rabbits. A curative effect was found.

VIII Curative effect of an agent at low dilution that uses to cause symptoms in a healthy organism , on a diseased or ill organism if the latter shows identical symptoms

This type of human study was used by Hahnemann in the early phase of detection, exploration and first clinical evaluation of 'homoeo'-pathy. See the contributions to clinical research.

IX Curative effect of an agent at high dilution (that, at low dilution, uses to cause symptoms in a healthy organism), on an organism intoxicated by a different agent at low dilution if the latter shows identical symptoms

In vitro study

The aim of these experiments is to study the specificity of the similia-principle. The most effective low-concentration (= relatively high dilution) stimulation of recovery processes in damaged cells was shown to occur when a high similarity was found between the damaged condition and the effect of the low dilution of the 2nd stimulus (van Wijk and Wiegant 1994). These observations confirmed the aspect of specificity of the similia-principle.

X Curative effect of an agent at high dilution in potentised form (that, at low dilution, uses to cause symptoms in a healthy organism), on an organism sensitised by a different agent at low dilution if the latter shows identical symptoms

These studies are related to (IX). In contrast to (IX), they concern not only the similarity principle in its complex form, but also the potency principle.

In vitro study

This type of study is closely linked to some immunological studies, i.e. where basophils are pretreated with antigens: Poitevin et al. 1988 studied the influence of traditional homoeopathic remedies on anti IgE antibody - treated basophils. An inhibition of basophil degranulation was found (##).

Zoological studies

Niebauer et al. 1979 studied the influence of a traditional homoeopathic remedy on experimental edema in rats. A curative effect was found. Bildet et al. 1989 studied the influence of potentised bee venom on UV-induced cutaneous erythema in guinea pigs (both a bee sting as well as ulraviolet rays can induce an inflammatory reaction). In this successively repeated study a curative effect was found (##).

Practical homoeopathy claims that homoeopathic dilutions of bee venom are capable of inhibiting cutaneous erythema induced both by bee venom as well as by ultraviolet rays.

Botanical studies

Speciani 1988 (thesis) performed several related botanical studies, with positive results.

XI Examples from modern conventional medicine

Mercury salts, as part of their toxicologic picture, cause oliguria and anuria; many of the classical diuretic drugs are mercury derivatives (e.g. mersalyl and chlormerodrin); the same holds true for the diuretics of the sulfonamide family (chlorthalidone, mefrusid, and chlorthiazide); digitalis, which in toxic quantities causes a picture of arrhythmic tachycardia, is indicated for the treatment of atrial fibrillation; at a high dosage, emetin causes a picture of diarrhoea and ulcers of the colon, resembling amebic colitis; in conventional medicine it is used as a drug against amebiasis (Oberbaum & Cambar 1994)

XII Curative effect of an agent at high dilution in potentised form (that, at frequent application at high dilution in potentised form, uses to cause symptoms in a healthy organism), on a diseased organism, if the latter shows identical symptoms

This type of treatment marks the specific frame of late Hahnemannian and modern homoeopathic medicine that is based α) on the similarity between the symptom picture (see V) of the agent and the symptom picture of the disease (*similarity principle*) and β) on the use of the agent at an adequate high dilution in potentised form (*potency principle*).

In vitro study

Poitevin et al. 1986 studied the influence of traditional homoeopathic remedies on basophils of hypersensitive (allergic) persons. An inhibition of basophil degranulation was found (##).

Zoological studies

Bildet et al. 1989 studied the influence of a traditional homoeopathic remedy on UV-irritated guinea pigs. This was done because of the similarity of the respective symptom pictures. A curative effect was found (see also XI) (##).

Khuda-Bukhsh and Banik 1990 studied the influence of a traditional homoeopathic remedy on damages in X-irritated mice. This was done due to the similarity of the respective symptom pictures. A curative effect was found (##).

Oberbaum et al. 1992 studied the influence of a traditional homoeopathic remedy on wound healing in mice. This was done because of the similarity of the respective symptom pictures. A curative effect was found.

Veterinary studies

Wolter 1985 studied the influence of a traditional homoeopathic remedy on labour pain in pigs. This was done because of the similarity of the respective symptom pictures. A curative effect was found.

Further veterinary studies are discussed in Kowalski 11989; Schütte 1994.

Human studies

are discussed in the contributions to clinical research.

Biophysical explanation

Ludwig 1987 and Smith 1987, 1994 described typical resonance frequencies related to potencies prepared according to the instructions of homoeopathy. At each step of the stepwise dilution and agitation process, more typical frequencies use to occur. This leads to the idea of an energetic interrelationship of a potency and an organism, possibly including positive and negative resonance phenomena. (Popp 1994, Bischof 1996).

Van Wijk and Wiegant 1989, 1994 and Endler et al. 1994a,b, 1995a,b studied the influence of potencies versus controls that were sealed in hardglass vials in order to make a direct mechanical contact between the dilution and the organism impossible. In these mutually comparable studies, significant effects were found on humans and on other vertebrates. The study of 1995b was performed multicentrically (##).

Senekowitsch et al. 1995 studied the influence of frequency patterns that were electronically scanned from a potency versus a control and stored on data carrier (CD). A significant effect, comparable to that of the original potency, was found (#).

Citro et al. 1991, 1994, 1995; Aissa et al. 1993a,b; Endler et al. 1995a,b; Benveniste et al. 1994; Thomas et al. 1995 studied the influence of test liquids that were prepared by means of electronic transmission of information from mother substances. In these mutually comparable studies, significant effects, compared to analogously prepared control, were found (#; 1995b ##).

Physics

Studies on physical properties of potencies (aqueous or on lactose) are discussed in Resch and Gutmann 1986; Luu-d-Vingh 1974 (##), Larue 1986 (##); Righetti 1988; King 1988, Majerus 1990, Weingärtner 1992; Antonchenko et al. 1992 Demangeat et al. 1992 (##), 1996; Bischof and Rohner 1992; Anagnostatos 1994; Endler and Schulte 1994; Endler et al. 1996, and Schulte and Endler this volume.

Recent physics research revealed that water dipoles may develop phase coherent oscillations through radiation coupling (Del Giudice et al. 1988, Del Giudice 1994). It is proposed that these could be modulated as a time-ordered pattern of signals and induce a coherent wave propagation (electron propagation). Further theory suggests that the phase coherent oscillations may originate information pattern formation through isotopicity effects in the dilutions (Berezin 1994, Schulte 1994).

Matter, energy and information. Based on previous attempts by Kaivarainen a mesoscopic attempt to describe solids and liquids has been brought into ultra high dilution context by Schulte recently (Schulte 1996) and has been further elaborated by

Schulte in this volume. This attempt is based on the de Broglie wave ansatz of matter which associates with every moving object of some momentum a well defined electromagnetic wave. Del Giudice und Preparata (1989, 1994, and this volume) showed that in solids as well as in liquids stable electromagnetic domains may exist. Theories like those based on fundamental concepts of physics will contribute to unravel the basic mechanism in homoeopathy and will give fundamental research into ultra high dilution and homoeopathy a solid base.

Conclusions and Implications to Clinical Research Issues

The presented classification shows a multitude of blind placebo-controlled studies as well as several meta-analyses, most of which have been published in conventional journals. Considering their international origin the results obtained appear too homogeneous to be purely coincidental. However, some of the more important studies have *failed* to confirm earlier results. For this reason it seems appropriate to speak of *clear indications rather than of proof of the correctness of certain homoeopathic concepts.*

A rough overall assessment of the accomplished work shows that both in biological and clinical experiments the effects of homoeopathic agents as compared with their matched controls (placebo) are only slight. If, for example, only studies complying to high methodological standards are considered, then the global difference in effect is around or below 10%. This applies both to published biological studies and to clinical studies (H. Walach); van Haselen et al. reported a global difference of 15%.

This overall outcome may speak in favour of the homoeopathic concepts under study, but even in cases where findings are backed up by statistically significant differences, they are nowhere near as convincing as the successes reported of homoeopathy in practice. Various explanations have been offered to account for this discrepancy:

- *So-called "research results" are purely coincidental.*
- *The successes reported by homoeopaths only refer to a few isolated cases.*
- *Fundamental research is irrelevant to homoeopathy.*
- *Affirmative results are usually not reproducible, that is, in the best case they are isolated incidents of some credibility; in the worst, they are attributable to inadvertent or even wilful distortion.*

In our opinion none of the above arguments finds support in the material we have compiled in this paper. However, a review of the presented results, both clinical and non-clinical, does suggest that even the most meticulous research on homoeopathy is bound to meet with greater uncertainties than conventional scientific research is, and that there may still be a number of unaccounted factors of influence. This is probably due to the tendency of homoeopathic potencies to produce true subtle effects which are seem to be very difficult to detect and to verify. Furthermore, the decisive role of complex, non-linear stimulus-response patterns in homoeopathic effects add to the difficulties in the detection mechanism.

Selection of material or patients/test subjects to be examined

Clinical homoeopathic research requires that certain fundamental principles are observed: the principle of *individualisation*, the choice of an *appropriate symptomatology*, and a certain level of *vital resources* still present in the patient. These requisites have been consistently observed in recent clinical studies.

Furthermore, experimental proof of homoeopathic effects presupposes the use of a sufficiently sensitive model. In fundamental research this limits the range of organisms suitable for research (e.g., organisms under stress); in clinical research, too, it may require a certain degree of sensitivity in the patient or test subject, similar perhaps to that of electromagnetically sensitive patients described in the conventional medical literature. According to Smith it is likely that only 10-20% of the population, depending on the accompanying conditions, will be sensitive to high homoeopathic potencies.

Drug proving, the actual empirical basis of homoeopathy, is probably subject to the same kind of uncertainty. In our view the current practice of testing homoeopathic drugs by provoking proving symptoms in healthy test subjects selected at random should be replaced by the more goal-directed strategy of conducting preliminary screening experiments to select particularly sensitive subjects. Some authors in the older homoeopathic literature endorsed the idea of restricting drug proving to sensitive subjects (e.g., the Materia Medica of T.F. Allen), a notion which certainly deserves thorough investigation. We strongly recommend the Eliminating the bond between patient and therapist through double blinding

Nevertheless, it is only because homoeopathy has begun to subject itself to the criteria of conventional research that our understanding of homoeopathy today has advanced from where it was a few decades ago. At that time the scientific standards of homoeopathic research were correctly rated as poor, in contrast to many other fields of scientific research. Thus, although the results we have to offer today may still be far from conclusive, we can confidently say that our methods have become more reliable and our statements better founded.

Acknowledgements

The two main lines of research on homoeopathy, namely *the test substance applied (potency research)* and *the reactional cybernetics of living systems (research on the similarity principle)* are treated in the mutually complementary books Endler and Schulte (Endler and Schulte 1994, 1996). Extensive information on the basis of the similarity principle is provided in van Wijk and Wiegant 1994.

Previous literature surveys were presented by Albrecht 1990; Bastide 1989, 1991, 1996; Bellavite and Signorini 1992, 1995; Bornoroni 1992; Haidvogl 1990; Linde 1994; Majerus 1990; Poitevin 1990; Righetti 1991; Wagner et al. 1991, "Forschung im Dienste der Gesundheit" 1992.

The authors wish thank all who gave their advice for establishing the classification frame, especially M. Haidvogl, P. Andersch, R. van Wijk, M. Bastide and F.A. Popp; those who helped in monitoring the studies listed here, especially M. Righetti, H. Walach, P. Fisher, G. King, J. Hornung, M. van Wassenhoven, M. Gooch and the group "Forschung im Dienste der Gesundheit"; those who helped in classifying the studies, especially F.A.C. Wiegant and B. Poitevin; as well as H. Walach, C. Heckmann and T. van Asseldonk for their valuable comments and E. Lauppert for his secretarial work.

References

Aissa J., Litime M.H., Attias E., Allal A., Benveniste J. Transfer of molecular signals via electronic circuitry. FASEB J. 1993a;7:A602.

Aissa J., Litime MH., Attias E., Benveniste J. Molecular signalling at high dilution or by means of electronic circuitry. J Immunolog 1993b;150:146A.

Albrecht H, Franz G. (ed.) Naturheilverfahren, Zum Stand der Forschung. Berlin: Springer 1990.

Anagnostatos G.S. Small water clusters (clathrates) in the preparation process of homoeopathy. In: Ultra High Dilution, Endler P.C. und Schulte J. (Hrsg.). Kluwer Academic Publishers. Dordrecht 1994:121-128.

Andresen M. Zytologische und mitochondriale Effekte einer Intoxikation mit CC14 am Beispiel des Lebergewebes der Ratte. Einfluß von Phosphorus D6 und Phosphorus D30. Dissertation Hannover 1985.

Antonchenko V.YA., Ilyin V.V. Points at issue in the physics of water und homoeopathy. In: Omeomed 92, First international Congress, Proceedings Book, University of Urbino. Editrice Compositori 1992.

Aubin M., Berjon J.J., Bildet J., Gomez H., Larrue J., Quilichini R. Etude de l'activité hépatoprotectrice de Phosphorus sur des fragments de foies des rats adultes placés en culture organotypique sur milieu artificiel après intoxication par CCl_4 .Ann Hom Fra 1980;22,3,25-33.

Aubin M., Bildet J., Bergeon J.J., Dufoeur H. Etude histoenzymologique de l'action de diverses dilutions de Phosphorus sur l'hépatite toxique du rat au tetrachlorure de carbone. Ann Hom Fra 1979;21, 359-366.

Auquière JP., Moens P., Martin PL. Recherche de l'action de dilutions homéopathiques sur les végétaux. J Pharm Belg I=1981;36:303-320. II=1982;37:117-134.

Bastide M, Daurat V, Doucet-Jaboeuf M, Pèlegrin A, Dorfman P. Immunomodulatory activity of very low doses of thymulin in mice. Int J Immunotherapy 1987;3:191-200.

Bastide M, Doucet-Jaboeuf M, Daurat V. Activity und chronopharmacology of very low doses of physiological immune inducers. Immunol Today 1985;6:234-235.

Bastide M. Report on research on very low dose effects. Giri, Atelier Alpha Bleue, Paris 1989.

Bastide M. Signal und Images. Paris: Alpha Bleue 1991.

Bastide M. (ed.) Signal und Images. Bi.annual GIRI Congress in Montpellier, France (1993) and Jerusalem (1994). Dordrecht: Kluwer Academic Publishers 1996.

Bastide M in: Taddei-Ferretti C (ed.). High Dilution Effects on Cells and Integrated Systems. Singapore: World Scientific, 1996.

Baumans V., Bol C.J., Oude Luttikhuis W.M.T., Beynen A.C. Does Chelidonium 3x lower serum cholesterol? Brit.Hom.J. 1987;76,14-15

Bellavite P., Signorini A. Fondamenti teorici e sperimentali della medicina omeopatica. Palermo: Nuova Ipsa Editore 1992.

Bellavite P., Signorini A. Homeopathy - A frontier in medical science. North Atlantic Books 1995.

Belon P. Homoeopathy und Immunology. Vol. of Preceedings. Congress Liga Medicorum Homoeopathicorum Internationalis, Airlington (USA) 1987.

Benveniste J,Davenas E,Ducot B, Cornillet B, Poitevin B, Spira A. L'agitation de solutions hautement diluées n'induit pas d'activité biologique spécifique. CR Acad Sci 1991;312 II:461-466.

Benveniste J., Aissa J., Litime M.H., Tsangaris G.Th., Thomas Y. Transfer of the molecular signal by electronic amplification. FASEB J. 1994;8:A398 (Abs.).

Benveniste J., Arnoux B., Hadji L. Highly dilute antigen increases coronary flow of isolated heart from immunized guinea-pigs. FASEB J 1992;6:A1610.

Berezin A.A. Ultra high dilution effect und isotopic self-organisation. In: Ultra High Dilution, Endler P.C. und Schulte J. (eds.). Kluwer Academic Publishers. Dordrecht 1994:137-169.

Betti L., Brizzi M., Nani D. MD., Peruzzi M. A pilot statistical study with homoeopathic potencies of Arsenicum album in wheat germination as a simple model. Brit.Hom.J. 1994;83,195-201.

Bildet J., Aubin M., Baronnet S., Berjon J.J., Gomez H., Manlhiot J.L. Résistance de la cellule hépatique du rat après une intoxication infinitésimale au tétrachlorure de carbone. Hom Fra 1984a;72,175-181.

Bildet J., Bononi F., Gendre P., Aubin M., Demarque D., Quilichini R. Etude au microscope électronique de l'action de dilutions de Phosphorus 15CH sur l'hépatite toxique du rat. Ann Hom Fra 1977;19,209-219.

Bildet J., Guere J.M., Saurel J. Aubin M., Demarque D., Quilichini R. Etude de l'action préventive de différentes dilutions de Phosphorus sur l'hépatite toxique du rat. Hom Fra 1984b&c;72,199-209.

Bildet J., Guere J.M., Saurel J., Aubin M., Demarque D., Quilichini R. Etude de l'action de différentes dilutions homéopathiques de Phosphorus blanc sur l'hépatite toxique du rat. Hom. fr. 1975; 72: 199-204.

Bildet J., Guyot M., Bonini F., Grignon MF., Poitevin B., Quilichini R. Mise en évidence des effets de dilutions d'Apis Mellifica et d'Apium virus vis à vis de l'érythème provoqué par un rayonnement U.V. chez le cobaye. Annales Pharmaceutiques Françaises 1989;47:24-32.

Bildet J., Saurel M., Aubin M., Cashoursat L. Etude de l'action de diverses dilutions d'ammanitine phalloide sur l'hépatite expérimentale au tétrachlorure de carbone. Ann Hom Fra 1978;4,271-276.

Bischof M, Rohner F. In: ZDN und FFB (ed.). Dokumentation der besonderen Therapierichtungen und natürlichen Heilweisen in Europa. Essen: Verlag für Ganzheitsmedizin 1992.

Bischof M. Biophotonen. Das Licht in unseren Zellen. Verlag Zweitausendeins, Frankfurt M., 1995

Bischof M. Biophotonen. In Endler P.C., Schulte J. (eds.) Homöo- und Bioresonztherapie. Voraussetzungen und Grundlagenforschung. Verlag Maudrich, Wien-München-Bern 1996.

Boiron J., Abecassis J., Belon P., Cazin J.C., Gaborit J.L. The effects of Arsenicum album 7CH upon rats poisoned with arsenic-quantitative study und statistical value of the results. 1978;Vol of Proc 35th Hom Congress,1-18

Boiron J., Belon P. Effects de dilution hahnemanniennes d'Histaminum 7CH et d'Apis mellifica 7CH sur la dégranulation des basophiles de patients allergiques. 35. Congresss Liga Medicorum Homoeopathicorum Internationalis, Airlington (USA) 1982.

Boiron J., Cier A. Elimination provoqueé et specifité d'action des dilutions infinitésimales d'élements toxiques. Ann. hom. franc. 1962a;4:789-795.

Boiron J., Cier A. Recherches expérimentales d'une activité isopathique. Ann. hom. franc. 1962b;10:796-800.

Boiron J., Cier A., Vingert C. Effects de quelques facteurs physiques sur l'activité pharmacologique de dilutions infinitésimales. Ann. hom. franc. 1968;10:187-196.

Boiron J., Marin J. Action d'une 15CH de sulfate de cuivre sur la culture de Chlorella vulgaris. Ann Hom Fr 1971;13:539-549.

Boiron J., Marin J. Action de dilutions homéopathiques d'une substance sur la cinétique d'élimination de cette même substance au cours de la culture de grains préalablement intoxiqués. Ann Hom Fr 1967;9:121-130.

Boiron J., Marin J. Action de dilutions successives de $HgCl_2$ sur la respiration de coléoptiles de blé. Ann Hom Fra 1965a&b;7,259-264&635-638.

Boiron J., Zervouacki. Action de dilutions infinitésimales d'arséniate de sodium sur la respiration de coléoptiles de blé. Ann Hom Fr 1962;5:738-742.

Bonavida B.TNF und TNF Receptors: Structure, mechanism of action und role in disease und therapy. Drugs of Today, 1992, Vol.28

Bonavida B., Safrit JT., Miorimoto H. Reversal of drug resistance: Synergistic Anti-Tumor Cytotoxic Activity by combinantion treatment with drug und TNF or toxins. Cancer Therapy, 1993,75:169-176

Bornoroni C. Synergism of action between indolacetic acid (IAA) und diluted solutions of $CaCO_3$ on the growth of oat coleoptiles. Berlin J Res Hom 1992;1(4.5):275.

Bornoroni C. (ed.) Omeomed 1992- Proceedings Book. Editrice Compositori

Boyd W.E. Biochemical und Biological Evidence of the Activity of High Potencies. Brt Hom J 1954;44:7-44

Cal J. C., Catroux P., Dorfman P., Cambar J.Variations saisonnières de l'action du mercure et du platine en homéopathie expérimentale. GIRI 27 (Abs.) 1988a.

Cal J.C., Larue F., Dorian C., Guillemain J.,Dorfman P., Cambar J. Chronobiological approach of mercury-induced toxicity und of the protective effect of high dilutions of mercury against mercury-induced nephrotoxicity. Liver Cells und Drugs 1988b;164,481-485.

Cal JC, Larue F, Guillemain J, Cambar J. Chronological approach of protective effect of Mercurius corrosivus against mercury induced nephrotoxicity. Ann Rev Chronopharmacol 1986;3:99-103.

Cambar J., Desmouliere A., Cal J. C., Guillemain J. Influence de l'administration de dilutions infinitésimales de mercurius corrosivus sur la mortalité induite par le chlorure mercurique chez la souris. Bull Soc. Pharmacol. Bordeaux 1983b;122:30-38.

Cambar J., Desmouliere A., Cal J.C., Guillemain J. Mise en évidence de l'effet protecteur de dilutions homéopathiques de mercurius corrosivus vis-à-vis de la mortalité au chlourure mercurique chez la souris. Ann Hom Fra 1983a;5,6-12

Cazin J.C. Etude pharmacologique de dilutions hahnemanniennes sur la rétention et la mobilisation de l'arsénic chez le rat. 1986;In:Boiron J., Belon P., Hariveau E.:Recherche en Homéopathie, Fondation Française pour la recherche en homéopathie, 19-39.

Cazin Jc, Cazin M, Gaborit JL, Chaoui A, Boiron J, Belon P, Cherruault Y, Papapanayoutou C. A study of the effect of decimal und centesimal dilutions of arsenic on the retention und mobilization of arsenic in the rat. Human Toxicol 1987;6:315-320.

Chaoui A. Influence de certains facteurs physiques et chimiques sur l'activité de dilutions infinitésimales d'arsénic. 1988;Thesis Univ Lille Fac de Pharm.

Citro M. Vom Pharmakon zur Frequenz; Elektronischer Medikamententransfer. In: II. internat. Symp. Biokybernetische Medizin, Würzburg 1991.

Citro M., Smith C.W., Scott-Morley A., Pongratz W., Endler P.C. Transfer of information from molecules by means of electronic amplification - preliminary results. In: Ultra High Dilution, Endler P.C. und Schulte J. (eds.). Kluwer Academic Publishers. Dordrecht 1994:209-214.

Citro M, Endler PC, Pongratz W, Vinattieri C, Smith CW, Schulte J: Hormone effects by electronic transmission. FASEB J, 1995 (Abs. 12025).

Cristea A. A homoeopathic remedy and the isolated rat duodenum. Cybernetica (Namur, B) 1991; 2:85-93.

Daurat V., Dorfman P., Bastide M. Immunomodulatory activity of low doses ofinterferon alpha.beta in mice. Biomed & Pharmacother 1988a;42:197-206.

Daurat V., Dorfmann P., Tetau M., Bastide M. Modulation de la réponse immunitaire cellulaire de la souris par des doses infrapharmacologiques d'immunomédiateurs: thymuline, interferon-β, interleukine 2. Cahiers de Biothérapie 1988b,99:81-94.

Davenas E, Beauvais F, Amara J, Oberbaum M, Robinzon B, Miadonna A, Tedeschi A, Pomeranz B, Fortner P, Belon P, Sainte-Laudy J, Poitevin B, Benveniste J. Human basophil degranulation triggered by very dilute antiserum against IgE. Nature 1988;333:816-818.

Davenas E, Poitevin B, Benveniste J. Effect on mouse peritoneal macrophages of orally administrered very high dilutions of silica. Eur J Pharmacol 1987;135:313-319.

Del Guidice E. Is the « memory of water » a physical impossibility? In: Ultra High Dilution, Endler P.C. und Schulte J. (eds.). Kluwer Academic Publishers. Dordrecht 1994:117-119.

Del Guidice E., Preparata E. A collective approach to the dynamics of water. MITH 1989.10, Geneva: CERN-Library.

Del Guidice E., Preparata E., Vitellio G. Water as a free electric dipole laser. Physical Review Letters 1988;61:1085-1088.

Dittmann J, Selbach AC, Hentges A, Harisch G. Use of urate oxidase test system to characterize the effect of homoeopathic potencies and of equally concentrated conventional dilutions. Phem Pharmacol Lett 1994, 4:19-22.

Dittmann J, Hentges A, Harisch G. Homoeopathic potencies and equally concentrated conventional dilutions as inhibitors or stimulators of acid phosphatase from potato. Pharm Pharmacol Lett 1994,4 :40-43.

Demangeat J.L., Gries P., Poitevin B. Modification of 4 MHz N.M.R. water proton relaxation times in very high diluted aqueous solutions. In: Bastide (ed.) xy 1996

Demangeat JL, Demangeat C, Gries P, Poitevin B, Constantinesco N. Modifications des temps de relaxation RMN à 4MHz des protons du solvant dans les très hautes dilutions salines des silice.lactose. J Med Nucl Biophy.1992;16,2:135-145.

Dorfman P. Le végétal comme materiél d´étude des dilutions homéopathiques. Cahiers de Biotherapie 1993;122:22-29

Doucet-Jaboeuf M, Guillemain J, Piechaczyk M, Karouby Y, Bastide M. Evaluation de la dose limite d'action du facteur thymique sérique. CR Acad Sci 1982;295 III:283-287.

Doucet-Jaboeuf M, Pèlegrin A, Cot MC, Guillemain J, Bastide M. Seasonal variations on the humoral response in mice following administration of thymic hormones. Ann Rev Chronopharmacol. 1984;1:231-234.

Doucet-Jaboeuf M, Pèlegrin A, Cot MC, Guillemain J, Bastide M. Variations saisonnières de la reaction immunitaire humorale chez la souris après adminstration d'hormones thymiques. Cahiers de Biothérapie 1984;84:81-84.

Doucet-Jaboeuf M, Pèlegrin A, Sizes M, Guillemain J, Bastide M. Action of very low doses of biological immunomodulators on the humoral response in mice. Intern J Immunopharmacol 1985a;7:312.

Doucet-Jaboeuf M., Pelegrin A., Bastide M., Guillemain J., Tetau M. Etude chronopharmacologique d'immunomodulateurs utilisés à doses infinitésimales chez la souris. Cahiers de Biothérapie1985b;88,41-43.

Doutremepuich C., Aguejouf O., Pintigny D., Serillanges MN, de Sèze O. Thrombogenic properties of ultra-low-dose of acetylsalicylic acid in a vessel model od maser-induced thrombus formation. Thromb. Res. 1994;76:225-229.

Doutremepuich C., De Sèze O., Anne MC., Hariveau E., Quilichcii R. Platelet aggregation on the whole blood after administration of ultra low dosage Acetylsalicylic Acid in healthy volunteers. Thrombosis Research 1987;47:373-377.

Doutremepuich C., De Sèze O., Le Roy D., Lalanne M.C., Anne M.C. Aspirin at very low dosage in healthy volunteers:effects on bleeding time, platelet aggregation und coagulation. Haemostasis 1990;20,99-105.

Dutta AC. Plants´ resposes to high homoeopathic potencies in distilled water culture. ICCHOS News Letter II 1989;3:2-8.

Endler P.C., Pongratz W. Homoeopathic effect of a plant hormone? A preliminary report. The Berlin J. on Research in Homoeopathy 1991;1:148-150.

Endler P.C., Pongratz W., Smith C.W., Schulte J. Non-molecular information transfer from thyroxine to frogs with regard to 'homoeopathic´ toxicology. J Vet & Human Toxicol 1995;37 (3).

Endler P.C., Pongratz W., Smith C.W., Schulte J., Senekowitsch F., Citro M. Non-molecular information transfer from thyroxine to frogs. In: Bastide M. (ed.) Dordrecht: Kluwer Academic Publishers 1996, in press.

Endler P.C., Pongratz W., van Wijk R., Waltl K., Hilgers H., Brandmaier R. Transmission of hormone information by nonmolecular means. FASEB J. 1994b;8:A400 (Abs.).

Endler P.C., Pongratz W., van Wijk R., Wiegant FAC., Waltl K., Gehrer M., Hilgers H. A zoological example on ultra high dilution research, 1994a. In: Ultra High Dilution, Endler P.C. und Schulte J. (eds.). Kluwer Academic Publishers. Dordrecht 1994

Endler PC, Pongratz W, Kastberger G, Wiegant FAC, Schulte J. The effect of highly diluted agitated thyroxine on the climbing activity of frogs. Vet Hum Tox 1994d;36:56-59.

Endler P.C., Schulte J. (eds.). Ultra High Dilution. Physiology und Physics. Kluwer Academic Publishers, Dordrecht, 1994c.

Endler P.C., Schulte J. (Hrsg.) Homöo- und Bioresonztherapie. Voraussetzungen und Grundlagenforschung. Verlag Maudrich, Wien-München-Bern 1996.

Forschung im Dienste der Gesundheit (Projektträgerschaft). Unkonventionelle Medizinische Richtungen. Bestandsaufnahme zur Forschungssituation. Wirtschaftsverlag, Verlag für neue Wissenschaft, Bonn 1992.

Fisher P, House I, Belon P, Turner P. The influence of the homeopathic remedy plumbum met. on the excretion kinetics of lead in rats. Human Toxicol 1987;6:321-324.

Fisher P., Capel I.D. The treatment of experimental lead intoxication in rats by plumbum metallicum und penicillamine. Proceedings of the 35th Congress of the Liga Medicorum Homoeopathica Internationalis, London 1982; 320-332.

Gaborit J.L. Etude pharmacologique de la rétention et de la mobilisation de l'arsénic sous l'effet de doses infinitésimales d'arsénic. DEA Pharmacochimie Univ Lille II Fac de Pharm 1987.

Gardes E. Effet d'une dilution infinitésimale d'acide nalidixique sur l'élimination de cette même molécule chez l'homme sain. Diplôme Pharmacie. Annales Homéopathiques Françaises 1989;77(5):60.

Graviou E., Biron A. Action d'une 15CH de sulfate de cuivre sur Lepidium en germination. Ann Hom Fra 1971;13,531-538.

Guillemain J., Cal J.C., Desmoulière D.,Tetau M., Cambar J. Effet protecteur de dilutions homéopathiques de métaux néphrotoxiques vis-à-vis d'une intoxication mercurielle. Cahiers de Biothérapie 1984;21,27-34.

Hadji L., Arnoux B., Benveniste J. Effect of dilute histamine on coronary flow of guinea-pig isolated heart. FASEB J 1991;5:A1583.

Haidvogl M. Naturwissenschaftliche Grundlagen der Homöopathie. Der Kinderarzt 1990;120:75-84.

Harisch G., Andersen M., Kretschmer M. Aktivitätsänderung von Succinatdehydrogenase und Glutathionperoxidase nach Verabreichung von homoeopathisch aufbereiten Phosphorverdünnungen im Rattenlebermodell. Dt Apotheker Zeitung 1986;6:315-320.

Harisch G., Kretschmer M. Jenseits von Milligramm. Die Biochemie auf den Spuren der Homöopathie. Springer Verlag, Berlin 1990.

Harisch G., Kretschmer M. Läßt sich die Wirkung von Homöopathika im Zellstoffwechsel nachweisen? Therapeutikon 1988b;3:190-194.

Harisch G., Kretschmer M., von Kries U. Beitrag zum Histaminrelease aus peritonalen Mastzellen von männlichen Wistar-Ratten. Deutsche tierärztliche Wochenschrift 1987;94:515-516.

Harisch G., Kretschmer M. Effekte homöopathischer Präparationen im Zellstoffwechsel. In: Stacher A ed, Wiener Dialog für Ganzheitsmedizin. Wien:J+V Edition 1988a.

Harisch G., Kretschmer M. Smallest zinc quantities affect the histamine release from peritoneal mast cells of the rat. Experientia 1988b;44,761-762.

Harisch G., Mueller R., Meyer W. Beitrag zum Leberstoffwechsel der Ratte nach Tetrachlorkohlenstoff unter Einfluß von Nux vomica D6 bzw. Flor de Piedra D4. Allgemeine Homöopathische Zeitung 1984;229,5:190-199.

Herkovits, Perez-Coll C., Zeni S. Incremento en la toxicidad del Cd por empleo de soluciones preparadas con agitacion y microdosis de Cd. medicina. (Buenos Aires) 1989;49:477-480.

Herkovits, Perez-Coll C., Zeni W. Reduced toxic effect of Cd on bufo arenarum embryos by means of very diluted und stirred solutions of Cd. Communicationes Biologicas 1988;7:70-73.

Hirst SJ, Hayes NA, Burridge J, Pearce FL, Foreman JC (University College London - Pharmacology Department - Gower Street). Human basophil degranulation is not triggered by very dilute antiserum against human IgE. Nature 1993;366:525-7.

Ives G. Relative permittivity as a measure of homoeopathic potency effect: Negative results on repeating previous work. Br.Hom.J. 1983;72:65.

Jones RL., Jenkins MD. Comparison of wheat und yeast as in vitro models for investigating homoeopathic medicines. Br Hom J 1983:72:143-147.

Jones RL., Jenkins MD. Plant responses to homoeopathic remedies. Br Hom J 1981;70:120-128

Khuda-Bukhsh AR., Banik S. Assessment of Cytogenical Damages in X-irradiated Mice und Their Alterations by the Oral Administration of a potentized Homeopathic Drug, Ginseng -200. Perspectives in Cytology und genetics 1990:7.

Kaavaarinen A. Mesoscopic Theory of Matter und Interaction with Light. Principles.of Selforganisation in Ice, Water und Biosystems. Turun Yliopiston Offsetpaino, Turku (FN) 1992.

King G. Experimental investigations for the purpose of scientifical proving of the efficity of homoeopathic preparations - a literature review about publications from Englishspeaking countries - Thesis. Tierärztliche Hochschule Hannover 1988.

König K. Über die Wirkung extrem verdünnter ("homöopathischer") Metallsalzlösungen auf Entwicklung und Wachstum von Kaulquappen. Zschft ges exp Med 1927; 56:881-593.

Kolisko L. Physiologischer und physikalischer Nachweis der Wirksamkeit kleinster Entitäten bei sieben Metallen. Goetheanum Vererlag, Dornach, Schweiz 1926.

Kowalski M. Homöopathische Arzneimittelanwendungen in der veterinärmedizinischen Literatur. Thesis Veterinärmedizin FU Berlin 1989.

Labonia W. et al. Acao das doses minimas na protecao do enevenamento oficido de animals de laboratorio. Congress of Liga Medicorum Homoeopathicorum Internationalis, Rio de Janeiro 1986.

Lapp C., Wurmser L., Ney J. Mobilisation de l'Arsénic fixé chez le cobaye sous l'influence de doses infinitésimales d'arséniate de sodium. Thérapie 1955;10:625-638.

Larue F., Cal J.C., Guillemain J., Cambar J. Influence de la durée de prétraitement sur l'effet de mercurius corrosivus vis-à-vis de la toxicité induite par le chlorure mercurique chez la souris. Hom Fra 1986;5,275-281.

Larue F., Cal J.C., Guillemain J., Cambar J. Influence du facteur dilution sur l'effet de mercurius corrosivus vis-à-vis de la toxicité induite par le chlorure mercurique chez la souris. Hom Fra 1985;73,375-380.

Larue F., Cal J.C. Mise en évidence de l'effet protecteur de différentes dilutions de mercurius corrosivus. Cahiers de Biothérapie 1985;88,71-78.

Lasne Ph.Propriétés des solutions "homéopathiques", mesure de la relaxation magnétique T2, Thèse pour le diplôme de doctorat en Pharmacie,, Lyon, France, 27 Octobre 1986.

Lauppert E. Auswirkung von 'homöopathisch'zubereitetem Kupfersulfat auf das Wachstum von Weizenkeimlingen. Thesis, Graz 1995

Lauppert E, Endler C. Enhanced inversion effect of thyroxine 10e-10 - 10e-13 by agitation. Angenommen für: Taddei-Ferretti C (Hrsg.): High Dilution Effects on Cells und Integrated Systems. Singapore: World Scientific, 1996.

Lehner xy In: Stacher A (ed.). Ganzheitsmedizin. Zweiter Wiener Dialog. Facultas-Universitätsverlag, Wien 1991,

Lessell C.B. The infinitesimal dose - The scientific roots of homoepathy. Saffron Walden, CW Daniel 1994.

Linde K. Dosisabhängige Umkehreffekte. Eine differenzierte Literaturbetrachtung. Thesis Ludwig.Maximilian.Universität München 1991.

Linde K, Jonas WB, Melchart D, Worku F,Wagner H, Eitel F. Critical review und meta-analysis of serial agitated dilutions in experimental toxicology. Hum Exp Toxicol 1994;13:481-482.

Litime M.H., Aissa J., Benveniste J. Antigen signalling at high dilution. FASEB J. 1993;7:A602 (Abs.).

Ludwig W. Eigenresonanzen verschiedener Wasserformen. Erfahrungsheilkunde 1987;36(13).

Luu-d-Vingh C. Les dilutions homoeopathiques. Controle et étude par spectrographie Raman-Laser. Thesis University Montpellier 1974.

Majerus M. Kritische Begutachtung der wissenschaftlichen Beweisführung in der homöopathischen Grundlagenforschung. Gesamtbetrachtung der Arbeiten aus dem frankophonen Sprachraum. Thesis Tierärztliche Hochschule Hannover 1990.

Martius F. Das Arndt-Schulz Grundgesetz. Muench Med Wschr 1923; 70,31:1005-1006.

Mouriquand G., Cier A., Boiron J., Edel V., Chighizola R. Rétention et mobilisation de toxiques exogènes chez le pigeon. Ann. hom. franc. 1961;10:766-781.

Mouriquand G., Cier A., Boiron J., Edel V., Chighizola R. Speicherung und Mobilisierung exogener Gifte bei der Taube. Allgemeine Homöopathische Zeitung 1962;7:325-337.

Mouriquand G., Cier A., Boiron J., Edel V., Chighizola R. Rétention et mobilisation de toxiques exogènes chez le pigeon. Ann Hom Fra 1975;17,325-336.

Netien G., Graviou E., Marin M. Action de doses infinitésimales de sulfate de cuivre sur des plantes préalablement intoxiquées par cette substance. Ann. hom. franc. 1965;7:248-252.

Netien G., Graviou E. Modifications rythmologiques d'un matériel sain traité par une 15CH de sulfate de cuivre. Ann Hom Fr 1978;4:219-225.

Netien G., Gray G.H., Boiron J. Action d'une quinzième centesimale de sulfate de cuivre sur lepidium en germination. Ann Hom Fra 1972;14,287-296.

Netien G. Action de dilutions homéopathiques sur la respiration des coléoptiles de blé. Ann Hom Fr 1962;4:823-827.

Niebauer G.W., Kläring W.J., Dorcsi M. Die Wirkung von homöopathischem Apis D30 im experimentellen Entzündungsmodell. Documenta Homoeopathica 1979;2,343-352.

Noiret R., Glaude M. Atténuation du pouvoir germinatif des grains de froment traités par $CuSO_4$ en dilutions homéopathiques. Rev Belg Hom 1979;31:98-130.

Noiret R., Glaude M. Etude enzymatique du grain de froment ontoxiqué par du sulfate de cuivre et traité par différentes dilutions hahnemanniennes de la même substance. Rev Belg Hom 1976;IX(1):461-495.

Oberbaum M, Markovits R, Weissman Z, Kalinkevits A, Bentwich Z. Wound healing by homoepathic silica dilutions in mice. Harefuah 1992;123:79-82.

Oberbaum M., Cambar J. Hormesis: Dose-dependent reverse effects of low und very low doses. In: Ultra High Dilution, Endler P.C. und Schulte J. (Hrsg.). Kluwer Academic Publishers. Dordrecht 1994,5-18

Ovelgönne J.H., Souren J.E.M., Van Rijn J, Van Wijk R. und Wiegant F.A.C. Enhancement of the stress response by low concentrations of arsenite-pretreated H35 hepatoma cells. Toxicology und Applied Pharmacology 1995;132:146-155.

Ovelgönne JH., Bol AJWM, Hop WJC, van Wijk R. Mechanical agitation of very dilute antiserum against IgE has no effect on basophil staining properties. Experientia 1992;48 (5):504-508.

Paterson J. Report on Mustard Gas Experiments (Glasgow & London). J Am Inst Hom 1944;37:47-88.

Pelikan W., Unger G. The activity of potentized substances. Experiments on plant growth und statistical evaluation. Brit Hom J 1971;60(4),233-266.

Petit C., Belon P., Got R. Effect of Homeopathic Dilutions on Subcellular Enzymatic Activity. Human Toxicology 1989;8:125-129.

Poitevin B, Davenas E, Benveniste J. In vitro immunological degranulation of human basophils is modulated by Lung histamine und Apis mellifica. Brit J Pharmacol 1988;25:439-444.

Poitevin B. Introduction à l'homéopathie. Bases expérimentales et scientifiques. Hrsg. CEDH, 1990.

Poitevin B. Introduction à l'homeopathie. CEDH 1990.

Poitevin B. Mechanisms and action of homoeopathic medicines. Recent findings and hypothesis1: Physico-chemical mechanisms. Br. Hom. J. 1995;84:32-33.

Poitevin B., Aubin M., Benveniste J. Approche d'une analyse quantitative de l'effet d'Apis mellifica sur la dégranulation des basophiles humains in vitro. Innovation et Technologie en Biologie et Medicine 1986;7:64-68.

Pongratz W., Endler P.C. Reappraisal of a classical botanical experiment in ultra high dilution research. In: Ultra High Dilution, Endler P.C. und Schulte J. (Hrsg.). Kluwer Academic Publishers. Dordrecht 1994:121-128.

Pongratz W., Endler P.C., Poitevin B., Kartnig Th. Effect of extremely diluted plant hormone on cell culture. Proc. 1995 AAAS Ann. Meeting, Atlanta 1995.

Popp F.A. Ergebnisse eines Forschungsauftrages zum Wirksamkeitsnachweis der Homöopathie. Bericht an Bonn. Essen 1986.

Popp F.A. Some biophysical elements of homoeopathy. In: Ultra High Dilution, Endler P.C. und Schulte J. (eds.). Kluwer Academic Publishers. Dordrecht 1994:177-185.

Popp F.A. Some biophysical elements of homoeopathy. In: Ultra High Dilution, Endler P.C. und Schulte J. (eds.). Kluwer Academic Publishers. Dordrecht 1994,177-185

Preparata G. QED Coherence and Condensed Matter. World Scientific, Singapore, 1995.

Projetti ML., Guillemain J., Tetau M. Effets curatifs et préventifs de dilutions homéopathiques de sulfate de cuivre appliquées à des racines de lentilles pré- ou post-intoxiquées. Cahiers de Biothérapie 1985;88:21-27.

Reilly D., Taylor M.A., McSharry C., Aitchison T. Is homoeopathy a placebo response? Controlled trial of homoeopathic potency, with pollen in hayfever as model. The Lancet 1986;19:881-886.

Resch G., Gutmann V. Wissenschaftliche Grundlagen der Homöopathie. Organon-Verlag, Berg.Starnberger See 1986.

Righetti M. Forschung in der Homöopathie. Grundlagen, Problematik und Ergebnisse. Burgdorf-Verlag, Göttingen 1988.

Roth C. Literature Review und critical analysis on the topic of "In- und Detoxication Experiments in Homoeopathy. Berlin J Res Hom 1991;1:111-117.

Sainte-Laudy J, Belon Ph. Inhibition of human basophils. Agent Actions 1993;38:245-247.

Sainte-Laudy J. Inhibition of basophil degranulation with histamine dilutions. C.R. de la Reunion de la Fond. Franc. pour la Recherche en Homéopathie, Lyon 1989.

Sainte-Laudy J., Haynes D., Gerswin G. Inhibition effects of whole blood dilutions on basophil degranulation. Int J Immunother II 1986;3,247-250.

Santini R, Tessier M, Belon P. Effects of low dose treatment with cuprum 4CH on neostigmine digestive action in female mice. Drug Develop Res 1991;24:231-233.

Schulte J. Conservation of structure in acqueous ultra high dilutions. In: Ultra High Dilution, Endler P.C. und Schulte J. (eds.). Kluwer Academic Publishers. Dordrecht 1994:105-115.

Schulte J. Mesoscopic theories in physics. University of Technology Publishing. Sydney 1996.

Schulte J and Endler PC. Elemente einer Theorie zu Homöopathie und "Bioresonanz". In Endler PC and Schulte J (Hrsg.) Homöo- und Bioresonztherapie. Voraussetzungen und Grundlagenforschung. Verlag Maudrich, Wien-München-Bern 1996.

Schütte A. Ist Forschung in der Veterinärhomöopathie gerechtfertigt? Berlin Münch. Tierärztl. Wschr. 1994;107,229-236.

Senekowitsch F., Endler P.C., Pongratz W., Smith C.W. Hormone effects by CD record.replay. FASEB J. 1995;9 (Abs. 12161)

Smith CW. Coherent electromagnetic fields und bio-communication. In Popp FA, Warnke U, König HL, Peschka W (eds.). Electromagnetic Bio-Information. Urban & Schwarzenberg, Baltimore 1989.

Smith C.W., Endler P.C. Resonance phenomena of an ultra high dilution. In: Ultra High Dilution, Endler P.C. und Schulte J. (eds.). Kluwer Academic Publishers. Dordrecht 1994:121-128.

Smith CW. Measurements of the electromagnetic fields generated by biological systems. Neural Network World 1995; 5,819-829.

Souza Magro I.A. et al. Reducao da nefrotoxidade induzida por aminoglucosideos. 41.Liga Medicorum Homoeopathicorum Internationalis Congress, Rio de Janeiro 1986.

Speciani L.C. Omeopatia e agricoltura. Thesis Univ. Milano 1988.

Steffen W.A. Growth of yeast cultures as in vitro model for investigating homoeopathic medicines. Brit Hom J 1984;73,198-210.

Sukul N.C., Zaghlool H.A. Effect of two homoeopathic drugs, Agaricus muscarius und Nux vomica on the isolated ileum of rats. Sci. Cult. 1990;56:254-258.

Sukul NC., Bala SK., Bhattacharya B. Cataleptogenic effect prolonged by homoeopathic drugs. Psychopharmacology 1986;89:338-339.

Sur R.K., Samajdar K., Mitra S., Gole M.K., Chakrabarty B.N. Hepatoprotective action of potentized Lycopodium clavatum. Brit Hom J 1990;79,152-156.

Taddei-Ferretti C. (Hrsg.). High Dilution Effects in Cells and Integrated Systems. Singapore: World Scientific 1996.

Thomas Y., Schiff M., Litime M.H., Belkadi L., Benveniste J. Direct transmission to cells of a molecular signal (phorbol myristate acetate, PMA) via an electronic device. FASEB J. 1995;9 in press (Abs).

Van Wijk R., Wiegant F.A.C. Cultured mammalian cells in homeopathy research; the similia principle in self-recovery. Utrecht University, Utrecht 1994

Von Zglinicki T., Edwall C., Ostlund E., Lind B., Nordberg M., Ringertz N.R. und Wroblewski J. (1992). Very low cadmium concentrations stimulate DNA synthesis und cell growth. J.Cell.Sci. 103:1073-1081.

Wagner H., Jurcic K., Doenick A., Rosenhuber E., Behrens N. Die Beeinflussung der Phagozytosefähigkeit von Granulozyten durch homöopathische Arzneipräparate. Arzneim.-Forsch.Drug Res. 1986;36(II):1421-1425.

Wagner H., Pezzuto J., Melchart D., Linde K., Kreher B. Effects of very low doses in immunology. The Berlin Journal on Research in Homoeopathy 1991;1(3),164-165.

Weingärtner O. Experimentelle Studien zur physikalischen Struktur homöopathischer Potenzen. In:Albrecht H.Franz G (eds). Naturheilverfahren,Zum Stand der Forschung. Berlin:Springer 1990.

Weaver JC., Astumian RD. The response of living cells to very weak electric fields. the thermal noise limit Science 1990;247:459-462.

Weingärtner O. Homöopathische Potenzen. Wunsch und Wirklichkeit bei der Suche nach der therapeutisch wirksamen Komponente. Berlin:Springer 1992.

Weisman Z., Topper R., Oberbaum M., Bentwitch Z. Immunomodulation of specific immune response to KLH by high dilution of antigen. Abstract GIRI Meeting Paris 1992a.

Weisman Z., Topper R., Oberbaum M., Harpaz N., Bentwitch Z. Extremely low doses of antigen can modulate the immune response. Proc VIII Intern Congress Immunol Budapest Hungary 1992b, 23-28 August, pp532.

Wijk van R., Wiegant F.A. Homoeopathic remedies und pressure-induced changes in the galvanic resistance of the skin. State University Utrecht 1989.

Wolter H. Wirksamkeitsnachweis von Caulophyllum D30 bei der Wehenschwäche des Schweins. In: Beweisbare Homöopathie. Gebhardt KH ed., 2 ed, Heidelberg: Haug Verlag 1985.

Wurmser L. Influence des doses infinitésimales sur la cinétique des éliminations. Hom.fr.1984;72:165-173.

Youbicier-Simo BL., Boudard F., Makaouche M., Bastide M., Baylé JD. Effects of embryonic bursectomy and in ovo administration of highly diluted bursin on adrenocorticotropic und immune responses in chickens. Int J Immunotherapy 1993;9:169-180.

STIMULATION OF CELLULAR DEFENCE AND RECOVERY BY SUBHARMFUL DOSES OF TOXICANTS THE HOMOLOGOUS COMPONENT OF THE SIMILIA PRINCIPLE

R. Van Wijk and F.A.C. Wiegant

The essence of homoeopathy is considered to be the stimulation of recovery and healing processes by application of compounds according to the similia principle. A research program was developed to study this fundamental principle in biological research at the cellular level. This research program has been subdivided in a number of sequential phases. In this chapter an overview is given of the so-called iso-pathic or homologous approach, which represents the similia principle in its most elementary form.

Mammalian cells were brought in a disordered state by exposures to damaging physical (heat shock) or chemical (sodium arsenite and cadmium chloride) conditions. In these dynamic states of disorder, defence and recovery processes are induced. At the cellular level self-defence and -recovery is largely dependent on the availability of so-called 'protector proteins'. Immediately following the induction of disorder, a low dose of the compound is then applied to a cell culture which has previously been disturbed by a larger dose of the same compound. Under the appropriate conditions production of protector proteins as well as development of tolerance can be stimulated to a larger extent in disordered cultures by application of low doses of the same condition in comparison with cell cultures which only received the disordering treatment without subsequent application of the low dose. The observed stimulatory effect appears to be specific for damaged and recovering cells as low doses of the initial agent do not exert any influence on undamaged control cells.

INTRODUCTION

The similia principle is considered to be the fundamental basis of homoeopathy. This principle pertains that disturbances (diseases) should be corrected by minute doses of any remedy, which at higher doses produces effects closely resembling the symptoms of the disease being treated. The most elementary approach of this curative strategy is the stimulation of self-defence and self-recovery processes by minute amounts of remedies which actually caused the disease or initiated the disturbed state. This last

J. Schulte and P.C. Endler (eds.),
Fundamental Research in Ultra High Dilutionand Homoeopathy, 215-228.

statement seems to be paradoxical. It poses the question if and, then so, how a minute amount of a toxic agent can contribute to the recovery after damage induced by a higher dose.

The similia principle can not be readily studied in one single type of experiment, since multiple aspects have to be analyzed and demonstrated either in a parallel or in a sequential manner. Therefore, a research program has been developed using a biological model system that allows the systematic unravelling of the various aspects involved (Van Wijk and Wiegant, 1994, 1997) (Table 1).

Table 1: Research Program on the Similia principle

Similia principle: Remedies (or conditions) that cause particular symptoms when applied to healthy biological systems are used to stimulate self-recovery in diseased biological systems showing the same symptoms.

In the analysis of the similia principle at the cellular level, the following three steps can be discerned:

Step 1 deals with the parameters for self-defence and self-recovery of cells in their disordered state. These studies require the analysis of the cellular response towards various toxins. Since protector proteins exert an essential role in cellular self-defence and -recovery, the types of protector protein which are induced will be evaluated in terms of damage- and recovery-specificity. The outcome may be that under different conditions of disorder, qualitatively different patterns of protector proteins are induced. These studies can, at least in part, be considered as analogous to description of the symptoms induced by different compounds at higher system levels known as 'provings' or 'remedy pictures' within homoeopathy.

Step 2 deals with the similia principle in its most reduced form, by determining whether self-defence and recovery in disturbed cell cultures is stimulated by application of a low dose of the disturbing agent. This approach can be defined as the homologous component of the similia principle, also known as the iso-pathic approach since a low dose of the identical compound is used in the curative strategy. The outcome is the analysis of a stimulation of self defence and recovery of disordered cells by low doses which are without effect in undamaged control cells. Subsequently, the degree and the duration of this increased sensitivity of diseased cells to homologous stimulation is analyzed. The outcome is the effectivity of homologous stimulation in the time period following exposure to the damaging agent.

Step 3 is focussed on the specificity of low dose stimulation. The question to be answered is whether self-defence after exposure to toxicant A is only stimulated in a homologous way, i.e., by low doses of A, or whether it is also effectively stimulated in a heterologous way by low doses of other toxicants such as B and C. This heterologous component of the similia principle requires the results of steps 1 and 2. After determination of the degree of (non)similarity between the symptom pattern of a disordered system and the 'remedy picture' of the employed compounds, it can be ascertained whether the degree of stimulation of recovery by low doses of different heterologous compounds is related to the degree of similarity.
Furthermore, the temporal aspect of stimulation of recovery by low doses of heterologous agents, as related to the state of sensitivity of the disordered system, is analyzed. The outcome of these studies leads to information about the relation between the homologous approach, which represents the similia principle in its most reduced form, and the heterologous approach, which includes the full spectrum of remedy-application according to the similia principle.

A number of starting points are relevant for this research program:

- 'self-defence' and 'self-recovery' of cells take a central position as regulatory principles at which homoeopathic treatments are directed. To this end a mammalian cell type was selected as model for the study of the similia principle, since cellular disregulation and recovery is generally considered to be the basis for regaining optimal organ functioning as well as of healing at the organism level.

- to maintain the link with homoeopathic clinical practice a cellular model is used which is in a disordered or diseased state and in which recovery processes are initiated. The effect of low doses is then studied in terms of stimulation of recovery processes.

- the recovery processes in this cell model are evaluated in a quantitative manner in order to elucidate different aspects of the similia principle.

- as the similia principle should have a general validity a range of different compounds or circumstances is selected.

- with respect to the concentration and preparation of the compounds to be used in low doses in order to study their stimulatory action in cellular self-recovery, a large percentage of remedies used in homoeopathic practice is in the concentration range of milli to micro molarity. Furthermore, it should be noted that remedies used in homoeopathic practice are not always potentized in the same way or to the same degree. Therefore, we chose to use dilutions which were not submitted to a specific serial succussion process but were normally diluted and vortexed according to usual laboratory procedures.

To understand the defence and recovery processes of the cell it is important to study the mechanisms that are activated following infliction of damage and thus to learn how the cell copes with this situation. Different types of damage can be inflicted on either DNA, proteins or membranes, each eliciting different mechanisms of defence and repair. In our research we have chosen to study proteotoxicity or damage to proteins as they form the major part of the protoplasm and are involved in all kinds of catalytic activities. Proteotoxicity is a phenomenon of increasing interest in biomedical disciplines, since it not only occurs after heat shock or after ingestion of environmental pollutants as heavy metals, but also following ischemia, damage by free radicals (oxidative stress), hypertension, inflammation, etc.

A variety of cellular alterations are observable in cells exposed to these damaging (proteotoxic) conditions. Part of these alterations or symptoms are expressions of processes leading to cell death following irreversible injury, whereas other alterations are expressions of processes leading to defence and recovery. A large group of damaging agents causes the production of a set of proteins, which is involved in cellular defence and recovery mechanisms. Due to the original observation that they are strongly induced following exposure to enhanced temperatures, these proteins are usually named 'heat shock proteins' or 'hsps' after the historical cause of their stimulation. Alternatively, they are now also named after their function, such as 'chaperone' or 'protector proteins'. There are different families of these proteins (hsp27, hsp60, hsp70, hsp90 and hsp100-family).

To study the effects of application of compounds according to the similia principle on the processes of self-defence, the following unambiguously measurable parameters were selected: synthesis of protector proteins and development of resistance towards damaging conditions (Van Wijk and Wiegant, 1994).

For research purposes the similia principle has been divided in two main components; i.e. a homologous and a heterologous component.

The homologous component is the most elementary part of the similia principle, since the agent which is used in a low dose to cure has also been responsible in high dose for disturbing the biological system.

In the heterologous component different substances or drugs are used for cure and previous disturbance of the biological system. The question to be answered is whether the degree of stimulation of self-recovery by low doses of homologous or various heterologous compounds is related to the degree of similarity in their effect.

In the full analysis of the similia principle a number of sequential steps can be discerned (Table 1).

In this chapter an overview is given of the homologous or iso-pathic approach. Further aspects of the research program on the similia principle are discussed elsewhere (Van Wijk and Wiegant, 1997).

EXPERIMENTAL APPROACH

Cell cultures and step-down protocols

Reuber H35 rat hepatoma cells in monolayer cultures have been used as a biological model system for studying the similia principle. The effects of the homologous or isopathic approach have been studied according to the so-called step-down protocol (Figure 1). According to this protocol, application of a high dose of a toxic compound to a cell culture is followed immediately by incubation in a lower concentration of the toxic compound used initially. During these prolonged incubations we studied the molecular processes indicative of the stimulation of cellular defence mechanisms i.e. induction of protector proteins, as well as development of cellular resistance. The detailed procedures have been published elsewhere (Wiegant et al 1993, 1994; Van Wijk and Wiegant, 1994).

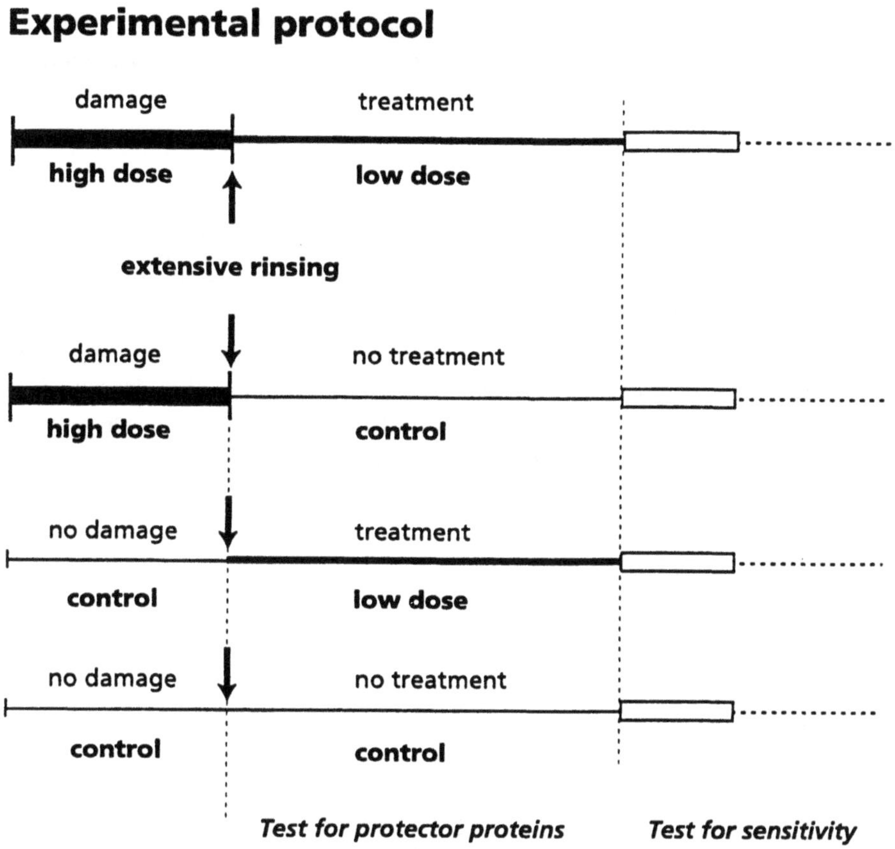

Figure 1. Schematic presentation of the step-down protocol (upper line) with cells receiving an initial high dose of a stressor followed by a prolonged incubation with a lower dose of the primary stressor and finally a test condition to evaluate development of tolerance. The other three lines represent the required control conditions.

Production of protector proteins

The rate of synthesis of protector proteins was determined by incubating cell cultures in medium containing labelled aminoacids (^{35}S)-methionine and (^{35}S)-cysteine). Cells were then solubilized and the newly synthesized and thus labelled proteins were separated by polyacrylamide gel-electrophoresis and assayed following autoradiography. The amounts of synthesized proteins were then determined by quantifying the intensities of the appropriate bands representing the different protector proteins.

Sensitivity to toxicants

A common method to establish cellular sensitivities for a toxicant is to evaluate the viability of the exposed cells, i.e. their reproductive potential in terms of their colony-forming ability. Cells which remain viable and functionally active are given the opportunity to divide and to form colonies during a period of one week. To this end the exposed cells are suspended by enzymatic dissociation and then inoculated in appropriate numbers into flasks containing growth medium. Since only viable cells will form a colony in the time period mentioned, the number of colonies observed after fixation and staining represent the survival ability.

RESULTS

Homologous approach

Introduction

It has been hypothesized that the similia principle can be tested at the cellular level in its most elementary form, by determining the extent to which recovery is stimulated by a small dose of the same substance which in first instance is responsible for upsetting the system or making it 'ill' (Van Wijk and Wiegant, 1994). This hypothesis of a beneficial effect of low doses of stress conditions has been studied using three different stressor conditions. Conditions were used that have the potential to cause cellular damage such as the physically damaging condition of increased temperature (heat shock)(Van Wijk et al., 1994) and chemical stressors such as sodium arsenite (Ovelgönne et al., 1995a; Wiegant et al., 1997a) and cadmium chloride (Wiegant et al., 1997b).

In our studies a heat shock was followed by a mild fever-like temperature, whereas a damage-inducing treatment with cadmium was followed by low subharmful doses of the same compound. The initially applied stressor conditions are known to exert proteotoxic properties at the high doses used and are thus effective inducers of various protector proteins in a variety of mammalian cells, including the cell line used in our studies (Wiegant et al., 1994, 1996; Ovelgönne et al., 1995b).

Arsenite treatments

With respect to cellular recovery from arsenite intoxications, our studies were aimed at the determination of the induction of mRNA's of protector proteins, the ensuing synthesis of protector proteins and the development of tolerance (Wiegant et al., 1993, 1997a; Ovelgönne et al., 1995a). . The question was asked whether these processes were sensitized to low concentrations of arsenite in arsenite pre-treated cells, and whether an ineffectivity of these low doses was observed in normal cells which did not receive the pre-treatment.

Using a step-down arsenite treatment (a 1 hour pretreatment with 100 or 300 μM arsenite followed by an incubation with lower concentrations (1-10 μM), cells were shown to exhibit increased sensitivities to these low concentrations of sodium arsenite with respect to their inductions of protector proteins and tolerance development. It was observed that under conditions of enhanced sensitivity an additional increase occurred in the synthesis of protector proteins as well as in their mRNA (as exemplified by hsp68 mRNA behavior) when low concentrations of arsenite were applied to these arsenite pretreated cells. Since no effects of these low concentrations were observable in unpretreated cells, the effects of step-down treatments thus results in higher effects than can be expected based on their summations.

Furthermore, it was observed that under these arsenite step-down conditions the development of tolerance was enhanced compared to the level of tolerance obtained by a single treatment with 100 or 300 μM arsenite for 1 hour (Ovelgönne et al., 1995a). When relatively high concentrations of arsenite were used in a pre-treatment the subsequent application either leads to further enhancement of tolerance (1μM) or to inhibition of tolerance development (10μM).

Heat treatment

Using step-down heating conditions a difference was also observed between mild and severe step-down conditions. Directly following an acute high-temperature heat treatment cells exhibit increased sensitivity to ensuing mild hyperthermic temperatures. In the first studies on the influence of step-down heating conditions an enhanced synthesis of protector proteins was demonstrated in the cells used in these studies (Schamhart et al., 1992). In additional experiments we determined whether the increase of the specific heat shock mRNAs following heat stress was also thermosensitized. To this end the induction of the mRNA of hsp68 was studied after step-down heating. The step-down heating procedure consisted of a pretreatment of 30 min at 41.5, 42.5 or 43.5°C, followed by continuous incubations at lower hyperthermic temperatures (40 or 41°C). Following mild pretreatments (41.5 and 42.5°C) the mRNA level of hsp68 was increased by subsequent incubation at 40°C, although incubation at 40°C alone exerted no effect. The increase was even more pronounced at 41°C post-treatment. In contrast, cell cultures pretreated with a severe heat shock (43.5°C) showed an inhibited or delayed synthesis of hsp mRNA when post-treated at the temperatures mentioned (Van Wijk et al., 1994).

In Figure 2 the effects of step-down heating (42.5°C followed by 39.5°C) on the production of various protector proteins is shown. It can be clearly observed that the synthesis of hsps is further enhanced during a prolonged period of time at 39.5°C).

Interestingly, an enhanced level of thermotolerance was also observed upon application of mild step-down heating. However, under conditions of maximal thermotolerance development a further stimulation of this development did not take place. These data imply that a shift in sensitivity occurs depending on the impact of the initial temperature of the step-down heating protocol (Van Wijk et al., 1994).

Cadmium treatment

Cadmium was selected as the third stressor condition to verify the mentioned hypothesis that the cadmium-induced cellular stress response can be modulated by the subsequent application of low concentrations of the same ion. It was shown that exposure of H35 cells to cadmium concentrations of 10 or 30µM for 1 hour leads to an initial sensitization towards a secondary treatment with this compound (Wiegant et al., 1997b). Furthermore, incubations for 1 hr with 10µM of cadmium induce a synthesis of the major protector proteins except for hsp60. A step-down cadmium regime, i.e. a pretreatment of 1 hr with 10 or 30 µM immediately followed by incubations with lower concentrations of cadmium (ranging from 0.03 to 1µM), leads to additional increases in synthesis of protector proteins. The sensitized cells do also develop higher levels of tolerance in the presence of the above-mentioned low concentrations of cadmium (Figure 3). No effects of these low concentrations could be observed on synthesis of hsps nor on tolerance development in non-pretreated cells.

In summary, using three different stressor conditions, we were able to show that a low dose of a stressor exerts a stimulatory effect on defence and recovery when applied to cell cultures which have been previously disturbed by a high dose of the same stressor.

In a next step, it is of interest to determine the specificity of the stimulatory effect of low doses of stressors. An overview of the results of this heterologous approach is described in Van Wijk and Wiegant (1997).

Desensitization towards homologous stimulation

A particular aspect of the mentioned stressors is, that as part of the recovery process, cells show a biphasic change in their sensitivity towards the stressor. An initially enhanced sensitivity (sensitization) towards a second application of the same stressor is followed by a reduced sensitivity (desensitization or tolerance) at a later time. The biphasic change in sensitivity was not only observed after heat treatments but also after treatments with arsenite (Wiegant et al., 1993, 1997a) or cadmium (Wiegant et al., 1997b). These changes in sensitivity are usually monitored as cell survival, but they also carry consequences for stressor-induced synthesis of protector proteins. Thus, it has been shown that cells exposed to an arsenite step-down protocol exhibit increased sensitivity towards low concentrations of sodium arsenite shortly following their exposure to the high concentration but not anymore when the lower concentration was applied 4 hr after the pretreatments (Ovelgönne et al., 1995a).

These observations lead to the conclusion that the capacity to stimulate recovery by an identical substance is a transient phenomenon. This has consequences for understanding the relation between the homologous and the heterologous approach as components of the similia principle.

One of the next questions in the research program is whether this desensitization in a later period of cellular recovery is a stressor-specific phenomenon and to which extent a heterologous induction remains possible.

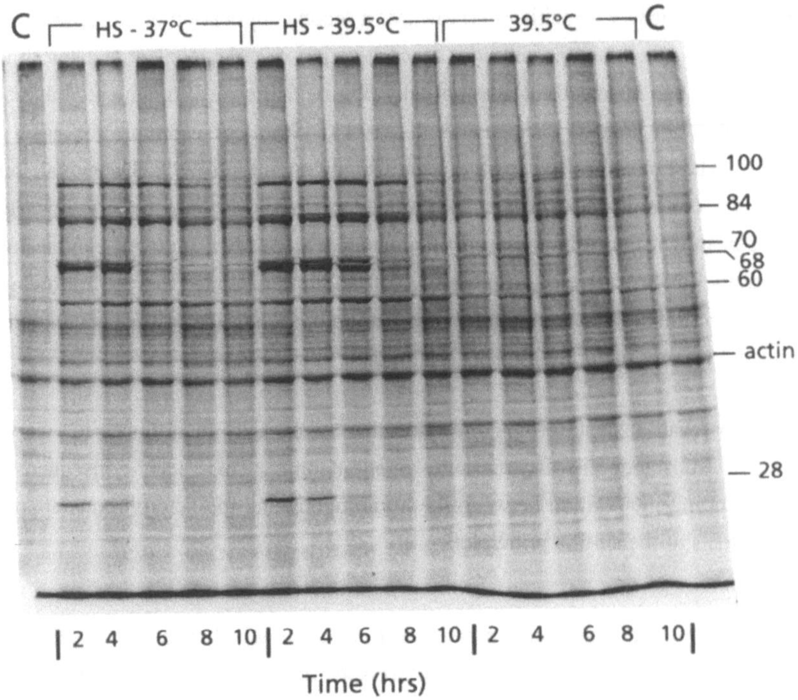

Figure 2. Induction of synthesis of protector proteins (hsps) in H35-cell cultures pre-treated with a heat shock (HS) of 42.5°C during 30 minutes, followed by a continuous treatment at the control temperature (HS-37°C) or by a mild hyperthermic temperature (HS-39.5°C) during 10 hours. At various times following the pretreatments, the cultures were incubated with (^{35}S)-labelled amino acids for 2 hour periods (0-2, 2-4, 4-6, 6-8 and 8-10 hours). Subsequently samples were prepared for one-dimensional gel-electrophoresis. Equal amounts of radioactivity was used in each lane. Shown are the polypeptide compositions of control cells kept at 37°C and of heat-treated cultures. The different hsps are indicated on the left with their molecular weight in kD.

Heterologous induction

The specificity of desensitization of protection protein induction was analyzed employing the three earlier mentioned damaging conditions (heat shock, sodium arsenite and cadmium chloride) as primary and secondary inducers of protection proteins. In both instances a high dose was applied (Wiegant et al., 1996). From the data it has been concluded that stimulation of protector protein synthesis following cadmium as the secondary stressor is severely inhibited in cadmium-pretreated cells, while re-inductions following arsenite and heat shock are specifically inhibited in arsenite- and heat shock-pretreated cells respectively. However, a less severe or hardly any inhibition was observed when a non-similar stressor was applied. The degree to which inhibition was observed did depend on the pretreatment. Interestingly, the pattern of (re-)induced protector proteins which was observed after application of the second stressor seemed to be characteristic for this secondary applied stress condition.

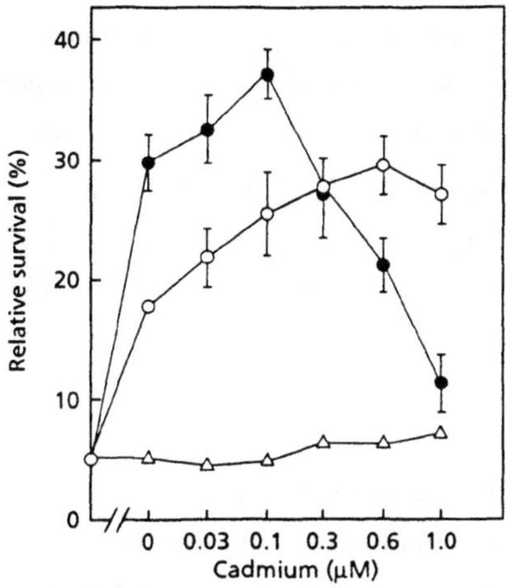

Figure 3. Level of tolerance induced by step-down cadmium treatments. Survival percentages are shown for cultures which were initially treated with 0 μM (Δ), 10 μM (O) or 30 μM (•) cadmium chloride, then with either 0, 0.03, 0.1, 0.3, 0.6 or 1.0 μM of this compound for 10 hours, and finally cell cultures were subjected to a test treatment with 50 μM cadmium for 2 hours. Results were corrected for pretreatment survival values.

In summary, homologous sensitization is transient and homologous desensitization develops specifically. Heterologous re-induction is still possible to occur but, in that case the pattern of protector proteins induced by the secondary applied stressor shows a stressor-specificity which is dependent on the second treatment only and independent of any pretreatment.

Future research aimed at further understanding of the similia principle

The possible implication of the results mentioned so far is that the increased production of protector proteins at later time following an insult, i.e. during the period of homologous tolerance, can only be obtained following application of a heterologous damaging compound. When we consider the pattern of synthesized protector proteins as the molecular symptoms of recovery, the specificity of the similia principle does reveal itself in the degree of relatedness between the effects of the disturbing compound on the pattern of synthesis of these protector proteins. It is therefore required to search for analogous damaging conditions which induce similar patterns of protector proteins as the primary stressor without being identical to this primary stressor. For this step 'provings' of different compounds have to be performed at the cellular level.

Another possibility is to look for conditions which do induce the whole range of protector proteins, such as for instance by the application of a mixture of various stressors in low doses. Then, as a result of a concerted action, enhancement of the synthesis of all protector proteins will take place due to an overruling or circumvention of the lack of induction of some protector proteins due to stressor-specificity.

In fact, from this perspective, nature seems to use fever (low hyperthermic) temperatures to provide organisms with a means to stimulate protection and recovery mechanisms following damage as was observed in preliminary data (Van Wijk and Wiegant, 1994).

CONCLUSION AND PERSPECTIVE

The aim of the presented overview is to illustrate the homologous component of the similia principle in an experimental framework. This approach is also known as the iso-pathic treatment since a disease or disturbance is treated with the causative agent of the disease, generally applied in lower amounts or under milder conditions. It can be concluded that during the transient period of enhanced sensitivity low doses of the original damaging agent are able to enhance the synthesis of various protector proteins and thus to induce higher levels of tolerance compared to control cultures that were sensitized but not incubated in the presence of low concentrations of the original stressor. Since these parameters reflect the self-defence and self-recovery response, and since we have observed this type of regulation employing three different damaging agents, our results suggest that this homologous or iso-pathic stimulation of recovery represents a general principle.

In this respect these observations not only seem to support the most elementary part of the age-old adage of 'like cures like', but also add some 'nuance' to these isopathic treatments. From our results at the cellular level it can be predicted that a beneficial effect of a low dose of a homologous agent can only be observed when:

- the concentration of a low dose post-treatment is carefully attuned to the severity of the disturbance
- the application is performed during the limited period of increased sensitivity following the disturbance.

Significance of fundamental studies

Studies on the similia principle at the cellular level are without doubt significant for our understanding of the stimulation of recovery processes at the cellular level. It is apparent that the regulation of recovery can be stimulated in a specific way by application of low doses of damaging conditions. The selection of conditions can be based on the molecular symptoms of the cellular defence process. In this respect, the application of low doses according to the similia principle definitely warrants further study.

It should now also be easier to understand the role of these cellular recovery processes for an organ's functionality after exposure of an intact organism to stressful conditions. Therefore it can be expected that the similia principle will manifest itself as a general biological phenomenon. This offers exciting opportunities for developing new avenues for therapeutic interventions in biomedicine.

Since our aim is to support the development of clinically useful agents or strategies to limit injury and promote functional recovery and/or repair, our results support the view that any manipulation directed at stimulation of defence and recovery responses following injuries will alter the final result. In this respect, when considering new approaches to rescue damaged cells, increases in cell 'defence capabilities' and stimulation of mechanisms leading to organism resilience, may be useful complementary approaches. In this respect, our results support the clinical use of low level exposures of stressors in order to stimulate cell recovery as performed in homoeopathy.

Acknowledgment

This work has been supported by the HomInt organisation. Part of this work has been published earlier in the HomInt Newsletter 97/1.

References

Ovelgönne JH, Wiegant FAC, Souren JEM, Van Rijn J and Van Wijk R. (1995a) Enhancement of the stress response by low concentrations of arsenite in arsenite-pretreated H35 hepatoma cells. Toxicol.Appl.Pharmacol 132: 146-155.

Ovelgönne JH, Bitorina M and Van Wijk R. Stressor-specific activation of heat shock genes in H35 rat hepatoma cells. (1995b). Toxicol.Appl.Pharmacol 135: 100-109.

Schamhart DHJ, Zoutewelle G, van Aken H and Van Wijk R. (1992). Effects on the expression of heat shock proteins by step-down heating and hypothermia in rat hepatoma cells with a different degree of heat sensitivity. Int.J.Hyperthermia 8: 701-716.

Van Wijk R, Ovelgönne JH, de Koning E, Jaarsveld K, Van Rijn J and Wiegant FAC. (1994). Mild step-down heating causes increased transcription levels of hsp68 and hsp84 mRNA and enhances thermotolerance development in Reuber H35 hepatoma cells. Int.J.Hyperthermia 10: 115-125.

Van Wijk R and Wiegant FAC. (1994). Cultured mammalian cells in homeopathy research; the similia principle in self-recovery. Utrecht: Utrecht University, pp 1-230.

Van Wijk R and Wiegant FAC. (1997). The similia principle in surviving stress; mammalian cells in homeopathy research. Utrecht University, Utrecht. pp 1-270.

Wiegant FAC, Souren JEM, Van Rijn J and Van Wijk R. (1993). Arsenite-induced sensitization and self-tolerance of Reuber H35 hepatoma cells. Cell.Biol.Toxicol. 49-59.

Wiegant FAC, Souren JEM, Van Rijn J and Van Wijk R. (1994). Stressor-specific induction of heat shock proteins in rat hepatoma cells. Toxicology 94: 143-159.

Wiegant FAC, Spieker N, Van der Mast and Van Wijk R. (1996). Is heat shock protein re-induction during tolerance related to the stressor-specific induction of heat shock proteins?. J.Cell.Physiol. 169: 364-372.

Wiegant FAC, Ovelgönne JH, Souren JEM and Van Wijk R. (1997a). Stimulation of self-recovery by low doses of arsenite in arsenite-intoxicated cells. In: Signals and Images (M. Bastide ed.). Kluwer Academic Publishers, Dordrecht. pp 41-51.

Wiegant FAC, Van Rijn J, Van Wijk R. (1997b). Enhancement of the stress response by minute amounts of cadmium in sensitized Reuber H35 hepatoma cells. Toxicology 116: 27-37

INFORMATION AND COMMUNICATION IN LIVING ORGANISMS

Madeleine Bastide

In many experiments, we could demonstrate the effectiveness of high dilutions prepared according to the homoeopathic process. Our experiments were performed by using serially diluted homoeopathic solutions of endogenous molecules active on the immune system. The results indicated that homoeopathic high dilutions of endogenous substances containing theoretically no molecule (> 10^{-24}) had an immunomodulatory and endocrine activities (Doucet-Jaboeuf *et al.*, 1982; Bastide *et al.*, 1985, 1987, 1995; Youbicier-Simo *et al.*, 1993, 1996a, 1996b, 1997). These activities were specific and the higher the dilution, the greater the capacity of the test substances to act on the organism. As a working hypothesis concerning the mechanisms underlying the observed effects of highly diluted endogenous compounds, we propose that non-molecular specific information corresponding to the original molecule is conveyed to the recipient organism which deals with this special cue to restore normal immune and endocrine functions.

We hypothesis that this information is likely transferred by electromagnetic waves. Information can be inhibited by strong electromagnetic fields as well as it can be transferred by an electronic device (Hadji *et al.*, 1992, Endler *et al.*,, 1994a, 1994b, 1997). Furthermore, this information is totally dependent on the receiver's state. In the endogenous molecule models, the organism recognizes self information automatically. These models are completely independent of the law of identity or the similia law which concern exogenous molecules and are helpful for the recognition of information by the organism.

As such, a new paradigm is necessary to attempt understanding the physical information of living organisms : just as the mechanistic paradigm seizes the material interactions between objects and the symbolic paradigm assumes linguistic facts, we introduce the paradigm of signifiers discovered by A. Lagache (Lagache 1988, 1997; Bastide & Lagache 1992, 1995, 1997).

J. Schulte and P.C. Endler (eds.),
Fundamental Research in Ultra High Dilutionand Homoeopathy, 229-239.

Effectiveness of high dilutions of endogenous molecule according to the receiver's state.

We first tested the effectiveness of high dilutions in mice. Classical immunopharmacological *ex vivo* models were used to evaluate the effect of diluted immunomodulators. Healthy mice pretreated with thymulin (a thymic hormone) 4C, 7C, 9C or 11C respectively, showed a significantly depressed specific humoral and cellular immune response with the highest dilutions. The effect on immunocompromised mice was a significantly stimulation of specific humoral and cellular immune responses (Doucet-Jaboeuf *et al.*, 1982; Bastide 1985, 1987, 1995). We always observed an opposite effect when we compared the results obtained in healthy mice with those obtained in immunocompromised mice (Bastide 1994).

We also tested the effect of bursin on the humoral immune response of bursectomized chickens (Youbicier-Simo 1993, 1996a, 1996b, 1997). Chicken embryos were surgically deprived of the Bursa of Fabricius at 80h of incubation and then treated in ovo with saline or bursin, (a tripeptide isolated from the bursa of Fabricius : Lys-His-Gly-NH_2), on days 6 and 9 of embryonic life. Four groups of chickens were tested : one untreated sham-operated control group ; one saline-treated bursectomized group ; one 7C bursin-treated group ; one 15C-20C (a pool of 15C to 20C dilutions) bursin-treated group. The four groups were repeatedly immunized with porcin thyroglobulin (Tg). Plasma corticosterone levels and serum titters of antibodies against thyroglobulin were evaluated. The immunized normal control group exhibited high concentrations of corticosterone and ACTH on day 38 while their bursectomized counterpart remained at baseline values. Normal hormonal levels were restored in the two high dilution-treated groups. Specific antibodies to Tg remained at background levels in the bursectomized group in spite of repeated stimulations, whereas the normal control group produced a strong and specific antibody response. *In ovo* administration of high dilutions of bursin induced high antibody production similar to that observed in normal chickens. These results were replicated successfully in other experiments including a randomized control tripeptide (Trp-Leu-Leu-NH_2) at the concentration of 100 fg. The latter tripeptide yielded no noticeable effect, therefore warranting the specificity of highly diluted bursin (Figure 1).

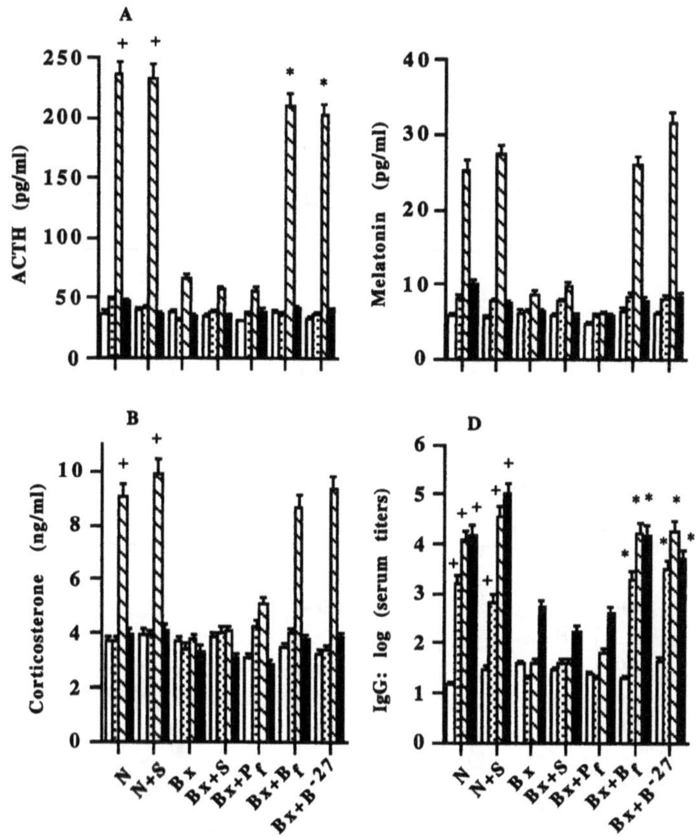

Figure 5: ACTH (A), corticosterone (B), melatonin (C) and anti-Tg IgG (D) responses to immunization against porcine Tg. Seven experimental groups were used (N, N+S, Bx, Bx+S, Bx+Pf, Bx+Bf, Bx+B^{-27}. The animals were immunized at the age of 21, 30 and 38 days. Blood samples were collected the day before the first immunization (d20: white columns) and at the age of 29 (dotted columns), 38 (hatched columns) and 47 (dark columns) to check the plasma levels of hormonal (ACTH,corticosterone, melatonin) and specific antibody (IgG) responses.
$^{+}$P<0.01 vs Bx; *P<0.01 vs Bx+S.

These results were emphasized by a study on the circadian rhythms of these hormones. The animals underwent a 12L-12D light-dark schedule for 12 weeks. Then, we measured the circadian rhythms of plasma ACTH, corticosterone (CORT), melatonin (MLT) and the pineal enzymatic activities of N-acetyl-transferase (NAT) and hydroxy-indole-o-methyl-transferase (HIOMT). The data showed that intact and bursin-supplied bursaless chickens exhibited pronounced diurnal rhythms of plasma ACTH and CORT, with low light-time levels and significantly higher night-time values reaching a sharp peak at midscotophase.

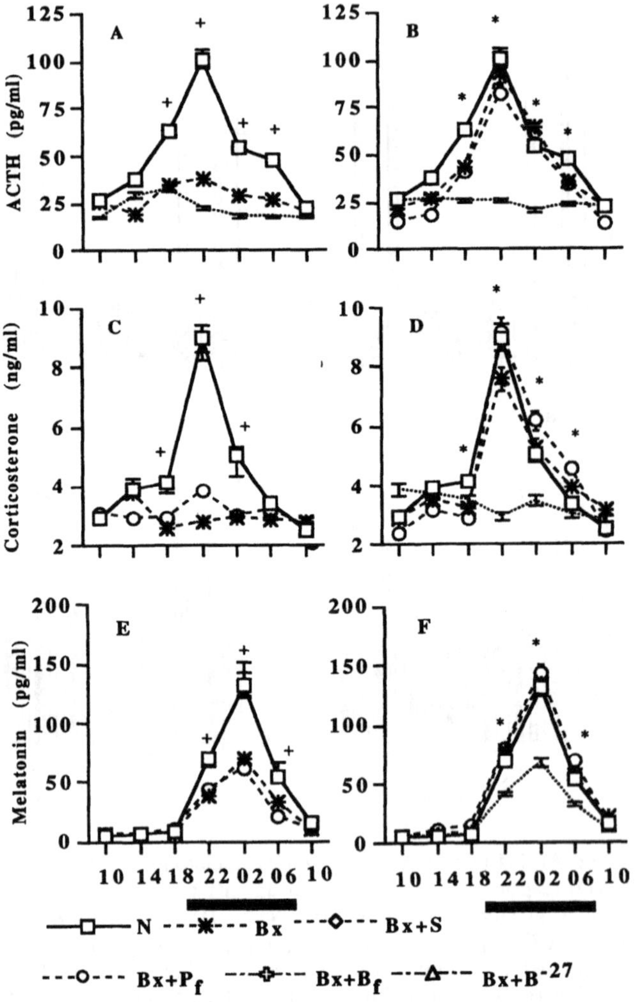

Figure 7. Circadian rhythms of the pituitary (A, B), adrenal (C, D) and pineal (E, F) hormones. The 7 groups of animals listed above were submitted to a 12L-12D alternating light-dark cycle, with lights off from 0700 to 1900 (black bar below the figures) .The plasma samples were assayed for ACTH (A, B), corticosterone (C, D) and melatonin (E, F). +P<0.01vs Bx; *P<0.01 vs Bx+S.

The waveform rhythms of both hormones were severed after bursectomy . In all groups and regardless of the treatment, a daily rhythm of MLT with middark crest was recorded and persisted in Bx birds despite 50% reduction in height. This waveform profile was in keeping with the rhythm of NAT, but HIOMT remained arhythmic. Neither the saline nor the control compound (Trp-Leu-Leu-NH_2) could reverse the effects of bursectomy.

The effect of administration of high dilutions of bursin / interleukin 3 / thymulin in total body irradiated mice gave us very precious indications on the activity of high dilutions (Guennoun *et al.*, 1996, 1997). The immunodepression model was a strong single total body irradiation of inbred BALB/c mice treated before and after irradiation in drinking water by high dilutions of a mixture of thymulin, bursin and IL-3. Our purpose was to evaluate the modifications in mortality rate of high dilution treated mice compared with irradiated untreated control mice. The evaluation of the recovery of lymphocyte status was also performed in some cases. The animals underwent total body irradiation (TBI) by ^{60}Cobalt γ- rays. The dose of irradiation was 6.3 ± 0.2 Gy (0.12 Gy/min); it was measured by PTW Unidos 0.125 cm^3 dosimeter. The beam covered the whole surface of the cage (20 cm x 20 cm) containing the mice. Mice were treated "per os" by drinking water. Before each experiment, all the solutions were blinded. When used, the solutions were succussed for 2min 30; then 2.5 ml were added to 250 ml of "Contrexeville" water (trade mark of a mineral water for consumers instead of tap water).The final solution was poured into the feeding bottle which was covered with a metallic paper (3 layers) and covered again with adhesive plastic (to avoid mice eating the metal). Mice were treated *ad libitum* starting 48 h before irradiation and throughout the experimentation. The different solutions were renewed every 7 days in the same bottles without any further sterilization. The following groups have been tested:

Control group: irradiated mice receiving only 2.5 ml of MilliQ pure water in the drinking water (USW).

Succussed solvent group: irradiated mice receiving 2.5 ml of succussed solvent diluted according to the remedies (SW pool 15-20C).

Succussed remedies: irradiated mice receiving the ternary highly diluted solution bursin-IL3-thymulin (BIT) (pool 15-20 C) in the drinking water (11 experiments throughout the year). When the mortality rate was expressed according to circannual variations, we observed that the radioprotective index ((Index = % mortality unsuccussed water group - % mortality Bursin/IL 3/THymulin) was higher in the cold season than in the warm season. These variations in sensitivity may be related to the circannual variations of the immune response. It was demonstrated that the cellular immune response was greater in the warm season while the humoral immune response had its acrophase in November-December (Mac Murray *et al*, 1983). The defence against immunodepression caused by irradiation is much more due to the cellular immune response than to the humoral response as in all acquired immune deficit. Taking this into account, it is evident that the irradiated mice were protected better in summer by high dilutions. Oppositely, they were not protected at all or died more than the untreated controls in winter as shown in Figure 3.

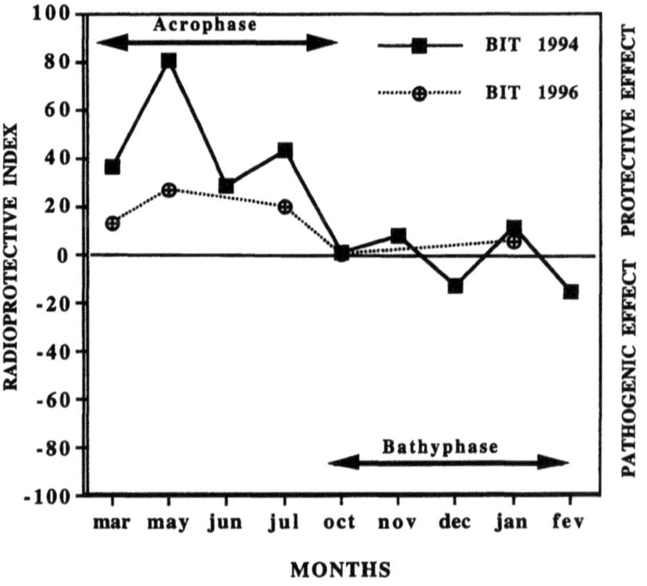

Figure 3: Highly diluted BIT radioprotective index according to the season (Guennoun *et al.*, 1996, 1997).

(Index= % mortality unsuccussed water group - % mortality Bursin/IL 3/THymulin)

An endogenous molecule information received and treated by a living organism is either corrective as a physiological response or pathogenic when it goes beyond the receiver's capability.

The experimental results reported here demonstrate the efficiency of highly diluted homoeopathic solutions of hormones or peptides in biological systems. Of course, all these experiments have nothing to do with the similia principle which is a communication system about symptoms. They raise the question of the nature of the thymulin-derived signal, for example at a concentration below 10^{-23} M, or bursin used at the theoretical concentration of 10^{-27}g (15-20 CH pool). These solutions are theoretically devoid of active molecules. We propose, therefore, that the succussed preparations obtained by successive 100-fold dilutions of immunomodulators transmitted "information" specific to the molecules used to prepare the solution, even though no active substance was present in the solution. This "information" has to be defined: this approach was permitted thanks to the new paradigm proposed by Lagache (Lagache 1988).

The paradigm of signifiers takes place within the framework of the logic of analogy. The communication will be mediated by *"semantic objects"* as physical signifiers : their application can be used to interpret the therapeutic effects of homoeopathic high dilutions. In fact, the living organism cannot be considered as an object: it functions as a whole and cannot be separated into its different elements; it is different before and after the events that it comes across; it is continuously modified as a function of time; it has a psychological memory and a physical one ; it is in a continual and irreversible learning process. The living organism (body and mind) is a complex structure able to exchange information with the external world; information also circulates in the internal world allowing exchanges at every level : *the living organism is an informed-informing structure. Living beings communicate with their world in a non verbal way, whether on a somatic or psychological level.*

The *semantic object* makes sense for the living organism at the body level : the circulation of this type of signifier has its own original laws.

- Each piece of information is not an object even if it has a *carrier*: for example, the information of the remedy is carried by the potentized dilution (the transmission of the information being due to the electromagnetic carrier obtained by the succussion of the solvent of the remedy as shown by Endler and colleagues (Endler *et al.* 1995, 1997).

- The sense is non-local; it concerns the whole organism designated as the receiver. This receiver then creates the meaning and is able to modify its behaviour.

- The simplest mode of representation for living structures is *a mimetic representation.* When the living organism receives this information brought by the semantic object (i.e. the potentized endogenous molecule) , it receives it not as a material object but as information about this object which calls for a processing and active regulation by the whole organism: this is an operation of *active mimesis* and the effect is compared to a physiological function. The bursectomized chickens treated by high dilutions of bursin is the best model: no residual information exist and a strong questioning of the embryos allows a good reception and treatment of the information which acts in a physiological manner.

However, when the mimetic representation is received but either too strong, or in such a way that the organism cannot react positively, the opposite effect is observed and designated as a *passive mimesis*: a pathological effect appears or increases. The model of the irradiated mice is helpful for understanding. It is a break-point model in which the slightest variation of the mice health state can induce either an active mimesis which gives a protective effect (acrophase of the cellular immune response during the warm season) or a passive mimesis which increases the pathogenic effect (lowest level of the cellular immune response in winter).

The same observation has been made by Endler et al. in the metamorphosis of the frogs modified by high dilutions of another endogenous molecule, thyroxine (Endler *et al.* 1997a, 1997b). As the frogs are normal and not deficient, the thyroxine information will be understood in different ways according to the capacity of the normal frog to treat this information. When the normal frog receives strong material doses (i.e. 10^{-9}M), the authors observe an acceleration of the metamorphosis of the normal frogs which can be considered as passive mimesis as a pathogenesis which would collect toxic effects. When a 1 to 100 dilution occurs, the active mimesis is observed and the metamorphosis is slowed down. When low potencies are used, the authors observe a slowing down effect; the treatment with high potencies shows a slowing down effect or an acceleration of the metamorphosis according to initial or experimental conditions: as we could demonstrate with the irradiated mice model, it is very easy to change the direction of the effect by using endogenous molecules: the information is treated but sometimes, it will produce the physiological effect (active mimesis) or the opposite effect (passive mimesis). The information received is an addition to the normal thyroxine information just at the period of the metamorphosis: this supplement of information can be misunderstood (as in our irradiated mice) and the opposite effect is observed.

These variations did not exist in the model of the bursectomized chickens: no bursin information was remaining as the chickens were surgically deprived at day 3 of the embryonic life. The highly diluted bursin was always acting in an active mimesis as the physiological effect because it was *replacing the normal information* of the molecule.

The similia law model

All these examples of the informative effect of high dilutions of endogenous substances are helpful to approach the functioning of information in living organisms. However, the effects of endogenous molecules is totally independent of the similia law or the identity law even though their interpretation can be done in the same paradigm. However, when high dilutions of an exogenous substance are administered to a patient, the framework of information which has to be used is the *similia law which consider only the symptoms*. In a patient, the pathological symptoms may be considered as an attempt at communication by the patient's body: the symptoms become the process of communication about the pathology, the organism being unable to find itself the mechanisms of regulation necessary to find again the normal state. These symptoms are read by the physician to identify the remedy able to give the whole symptoms in a healthy subject (similia law): the homoeopathic remedy will act as a semantic object and will transmit an artificial and alleviated information on the whole symptoms . Then, the body is able to recognize and treat this information which allows the negation of the symptoms recognized as an erroneous adaptation. The more similar the symptoms of the patient and the remedy, the more effective the remedy. The medical device has to re-inform the patient and makes his symptoms move on towards a higher level of integration. The action of the homoeopathic remedies consists of a dynamic analogy between pieces of information as proposed by A. Lagache in the paradigm of signifiers.

The observation of the biological effects of high dilutions of endogenous molecules such as hormones (thymulin, bursin, thyroxin) or cytokins have been of help for understanding the information process of a living organism: since these self-informations were naturally and automatically recognized, no framework of lecture such as identity law or similia law were necessary. As such, we could demonstrate the organism's capacity to increase a pathology or to correct it according to its capacity to treat the endogenous molecule information transmitted by potentized high dilutions.

References

Bastide M, Daurat V, Doucet-Jaboeuf M., Pelegrin A., Dorfman P., Immunomodulatory activity of very low doses of thymulin in mice, *Int.J.Immunotherapy*, 3: 191-200, 1987.

Bastide M. et Lagache A., *The Paradigme of Signifiers* , Alpha Bleue, Paris, 1992.

Bastide M., Immunological Examples on Ultra High dilution Research, in *Ultra High Dilution, Physiology and Physics,* Endler and Schulte Eds, Kluwer Academic Publisher, Dordrecht, 27-33, 1994a.

Bastide M., Boudard F., A novel concept of immunomodulation, in *Forum sur l'Immunomodulation,* M.Guenounou Ed, John Libbey Publishers Paris, 303-316, 1995.

Bastide M., Doucet-Jaboeuf M., Daurat V, Activity and chronopharmacology of very low doses of physiological immune inducers, *Immunol. Today,* 6: 234-235, 1985.

Bastide M., Lagache A. A new paradigm applied to high dilution effects on the living body. in: C Taddei and P Marotta eds: *High dilution effects on cells and integrated systems.* World Scientific Publ. Singapore, New Jersey, London, 1997.

Bastide M., Lagache A., Lemaire-Misonne C., Le paradigme des signifiants: schème d'information applicable à l'Immunologie et à l'Homéopathie, *Revue Intern. Systémique*, 9: 237-249, 1995.

Doucet-Jaboeuf M, J.Guillemain, M.Piechaczyk, Y.Karouby., M.Bastide, "Evaluation de la dose limite d'activité du Facteur Thymique Sérique", *C.R.Acad.Sci.*, 295, III 1982.

Endler P C, W.Pongratz, R.Van Wijk, F.A.C.Wiegant, K.Waltl, M.Gehrer, H.Hilgers, "A zoological example on ultra high dilution research. Energetic coupling between the dilution and the organism in a model of amphibia", in *Ultra High Dilution, Physiology and Physics,* Endler and Schulte Eds, Kluwer Academic Publisher, Dordrecht, 39-68, 1994a.

Endler P.C, W.Pongratz, G.Kastberger, F.A.C.Wiegant, J.Schulte, "The effect of highly diluted agitated thyroxine on the climbing activity of frogs", *J.Vet.Hum.Tox.*, 36, pp 56-59, 1994b.

Endler P.C, W.Pongratz, CW Smith, J.Schulte. "Non-molecular information transfer from thyroxine to frogs with regards to 'homeopathic' toxicology".*J.Vet.Hum.Tox.*, 37, pp 259-260, 1995.

Endler PC, W.Pongratz, C.W.Smith, J.Schulte, F.Senekowitsch, M.Citro, "Non molecular information transfer from thyroxine to frogs", in *Signal and Images*, M.Bastide Ed, Kluwer Academic Publisher, Dordrecht, 149-161, 1997a.

Endler PC, W. Pongratz, "Potentised thyroxine and Amphibian metamorphosis: does laboratory research reflect the natural laws which also form the basis of homeopathy ? R&D Newsletter, HomInt, 1997b.

Guennoun M, F Boudard, C Cabaner, Y Robbe, JB Dubois, M Bastide. "Radioprotection and immune system regeneration of irradiated mice by using high dilution treatment." *Alternative Therapies* , pp 64, January 1996, San Diego, USA

Guennoun M, F Boudard, C Cabaner, Y Robbe, JB Dubois, M Bastide. "Radioprotection and immune system regeneration of irradiated mice by using high dilution treatment." *Chronobiology, International* 14: 60 (Sup 1), 1997.

Hadji L., B.Arnoux and J.Benveniste, "Effect of dilute histamine on coronary flow of guinea-pig isolated heart. Inhibition by a magnetic field". *FASEB*, n°7040, 1992.

Lagache A., *Echos du Sensible.* Alpha Bleue, Paris, 1988.

Lagache A,"Notes on the conceptual basis of Science", *Signal and Images*, M.Bastide Ed, Kluwer Academic Publisher, Dordrecht, pp265-279, 1997a.

Lagache A,"What is Information", *Signal and Images*, M.Bastide Ed, Kluwer Academic Publisher, Dordrecht, 279-292, 1997b.

MacMurray J, Barker J, Armstrong J, Bozzetti L, Kuhn I. Circannual changes in immune functions. *Life. Sci.* , 32:2363-2370, 1983.

Youbicier-Simo B.J., Boudard F., Mekaouche M., Bastide M., Baylé J.D. (1993) Effects of embryonic bursectomy and *in ovo* administration of highly diluted bursin on adrenocorticotropic and immune response of chickens, *Int.J.Immunotherapy,* 9: 169-180.

Youbicier-Simo BJ, Boudard F, Guellati M, Mekaouche M, , Baylé JD, Bastide M.(1997), The role of the Bursa of Fabricius and highly dilute bursin in immunoendocrine interactions in the chickens. in *Signals and Images,* M.Bastide Ed., Kluwer Academic Publishers, Dordrecht, Boston, London, 111-121.

Youbicier-Simo BJ, Boudard F, Mekaouche M, , Baylé JD, Bastide M. (1996b. Specific abolition reversal of pituitary-adrenal activity and control of the humoral immunity in bursectomized chickens through highly dilute bursin. *Intern. J. Immunopathol. Pharmacol.*, 9: 43-51.

Youbicier-Simo BJ, Boudard F, Mekaouche M, , Baylé JD, Bastide M.,(1996a), A role for Bursa Fabricii and bursin in the ontogeny of the pineal biosynthetic activity in the chicken.*J.Pineal Res.*, 21: 35-43.

ARE THE CLINICAL EFFECTS OF HOMOEOPATHY PLACEBO EFFECTS?

REVIEWING THE EVIDENCE ON HOMOEOPATHY

Klaus Linde

Previous review of homoeopathy

I was always surprised and confused by the unscientific nature of the 'scientific' discussion of homoeopathy. In the main, it is not a discussion of the principles, characteristics and implications of homoeopathy but of the question "do potencies and in particular, high potencies, have a specific effect?" The translation of this question into clinical research leads to the question posed by David Reilly in the title of his famous hayfever trial published in *Lancet* in 1986: "Is homoeopathy a placebo response?" (Reilly et al. 1986).

There are many experts - both homoeopaths and scientists - who have drawn their "scientific" conclusions about the evidence for or against the existence of effects of high dilutions or high potencies. In the beginning, I was completely confused by the fact that everyone seemed to draw a different conclusion. Consider the following two examples:

The first is taken from a joint statement by Prokop, Glowatzki, Binder, Posnansky and Hopff in a book by the pharmacologist Hopff (Hopff 1991, *Homoeopathy - a critical view*, published by a major German mainstream medical publisher (Thieme)):

"It is important to mention that homoeopathy has often been tested by competent scientists of many disciplines in public institutions, as well as in special clinics at universities. It has however, never been shown to be superior than non-treatment or placebo."

The second example is taken from a review of the Austrian homoeopathy (Müller, 1991):

J. Schulte and P.C. Endler (eds.),
Fundamental Research in Ultra High Dilutionand Homoeopathy, 241-251.

"Summarising, it can be concluded that numerous laboratory and clinical experimental studies, conducted in strictly scientific conditions, prove clearly the efficacy of homoeopathic remedies."

After having read a number of such review papers on homoeopathy, I thought that the reason for these contradictory conclusions was that a small number of experts only cited results consistent with their previously held conclusions. A larger number whilst preferring to cite results concerning with their "conclusions". also tried to be as "objective" as possible, by discussing contradictory results. However, there were always weaknesses and deficiencies in these contradictory studies, and so, in the end, the original conclusion could be reached.

In 1991, a paper was published in the British Medical Joumal (Kleijnen 1991), which - as I thought - had surmounted all the drawbacks of these other reviews. Jos Kleijnen, Gerben ter Riet and Paul Knipschild from the Department of Clinical Epidemiology at the University of Maastrict had collected all available controlled clinical trials of "homoeopathy". Controlled clinical trials are seen as the best type of evidence for showing that an intervention is effective. Kleijnen et al. not only collected every trial they could but, in addition had taken a further very important further step: They had predefined their criteria to distinguish the "good" from the "bad", valid and invalid data. They used a system which gives points (a score) if certain methodological criteria are fulfilled and came up with charts of the trials of homoeopathy. The best trials were at the top of the charts, the worst trials were at the bottom. They found that nearly 80% of the trials results favourable for homoeopathy, and among the better trials about two thirds had "positive" results.

For convinced homoeopaths it was clear: -Homoeopathy was proven scientifically". For fully undecided people the conclusion was probably: "Homoeopathy might work". Kleijnen et al., who are open sceptics had a very interesting conclusion: they would b e ready to accept that homoeopathy can be effective if only the mechanisms of action could be determined. Peter Götzsche, a well-known methodologist from Denmark, commented on the review of Kleijnen et al. in a letter to the Lancet, in a way that probably represents the opinion and conclusions of the group of people whom I would call "convinced sceptics": "Rather than accepting the absurd idea that placebo can be better than itself, it seems more realistic to assume that the positive results in the homoeopathy trials were caused by bias." (Götzsche 1993).

There were two critical points in the review by Kleijnen and colleagues:

- The first was on the score for assessing the quality. In general, such scores are controversial as there is no gold standard to quantify quality. In particular, Kleijnen's score did not include a methodological key issue, the question of bias induced by excluding patients from the analysis who originally entered in the trial.

- The second critical point was the way they assessed the study results. Kleijnen et al. did a so-called vote-count they separated the trials into "positive" and "negative", mainly based on the conclusion of the authors of the original trials. This method is extremely crude and has problems regarding sensitivity (high likelihood of missing small effects), reliability (different authors may come to different conclusions), and validity (does the "vote" really represent the study result). From a homoeopathic viewpoint a lot of criticisms could have been raised, but as the results of the Kleijnen review were so favourable for them, homoeopaths were very reticent to speak out.

Our own systematic review

As even this excellent review had relevant shortcomings and the controversy on homoeopathy was still not solved, we thought arrogantly, perhaps!: We could do the definitive review of controlled clinical trials of homoeopathy, avoiding the shortcomings of the previous reviews and providing final answers. For more than three years we have worked on this review. It is a collaboration between a number of people in Munich (Nicola Clausius, Dieter Melchart, Florian Eitel), Wayne Jonas, the director of the NIH office of Alternative Medicine in Bethesda, Gil Ramirez, the Co-director of the San Antonio Cochrane Center, Texas, and Larry Hedges, Professor of Statistics in Chicago. Our protocol versions were peer reviewed by a number of critical and helpful experts including Iain Chalmers, Andy Oxman and Peter Götzsche.

Some days ago, we did the first analyses of all our data and the answer seems to be clear, it is " No", or in numbers: the odds ratio is 1.34.

What does this mean:

I do not know how many of you know Douglas Adams' comedy science fiction radio series 'The Hitchhikers Guide to the Galaxy' which was later published as a novel. There is a chapter in which Adams is narrating the story of a very intelligent population who developed a huge computer whose only task was to find the final answer to all questions. The computer thought and calculated for several million years and finally gave out the information that he had found the final answer. A number of scientists were elected to be the first to hear it. When the great moment had come the computer seemed to be very nervous and did not want to come up with it. But finally he told the scientists that the answer was 42. The scientists were really struck. The computer then told them that they now had to find the exact question for that answer. It would be unable to resolve the problem but it could plan a new and better Computer and after several billion years, it would get the question.

We had an explicit question in undertaking our review. However I am not sure if our answer is meaningful. You may consider it for yourself. What exactly did we do?

The primary question was:

Is there evidence that all homoeopathy (is) a placebo response? Or put another way: What is the mean effect of homoeopathy compared with placebo, and is this compatible with the placebo hypothesis? From the beginning we were aware that this question was problematic (would you ask: is drug therapy a placebo response?), but - as mentioned above - this is the crucial paint in the academic discussion of homoeopathy.

We had several sub-questions, but even though they are still broad and crude, they may be difficult to understand for those knowing little about homoeopathy:

Is classical homoeopathy a placebo? is clinical homoeopathy a placebo? Is complex homoeopathy a placebo? Is isopathy a placebo? are high potencies placebos?

We decided to use two approaches to answer our questions. In the first, all studies were grouped together, regardless of comparability in a clinical sense. In the second one, we looked for subsets of independently replicated clinically comparable trials, using similar homoeopathic interventions (e.g. the same remedy) in similar patients (e.g. pollinosis) measuring similar outcomes (e.g. severity of ocular symptoms).

Methods

We conducted an extensive literature search for prospective randomised, and/or double-blind, clinical trials comparing a homoeopathic intervention to a placebo. From the trials identified, we extracted information on the patients, the intentions, outcomes, methodological details and results. Methodological quality was rated using two different score systems, with a higher score was designating "better" quality.

Finally, the results of the clinical trials selected had to be summarised. We used a predefined strategy with preferences as to what, and how, to extract the results from the single studies. As a first preference we extracted data for predefined main outcome measures, second and third preferences were patient and physician global assessments, and if a study had no such data we choose the outcome which seemed clinically most relevant.

If results from several different studies are quantitatively summarised, the survey is called 'meta-analysis'. In doing this, first the size of the effect of every study is estimated and then all the material is lumped together and an average effect size is calculated. Normally, this should be done with studies which are similar with regard to patients, interventions, outcomes and quality. Even then, meta-analysis may be problematic. We had studies differing in all respect; but as our undifferentiated question required such an approach, we lumped everything together.

Additionally, we globally categorised a study result as positive (homoeopathy better than placebo), no difference, or negative (placebo better from homoeopathy). This method is called a vote count.

Results

The results presented here are preliminary results from our first analysis; the final results will be published elsewhere.

Ule identified 184 controlled clinical trails of homoeopathy. 35 investigated an individualised intervention ("classical" homoeopathy), 101 the use of a single remedy for a certain condition ("clinical" homoeopathy), 30 complex preparations (fixed combinations of remedies for certain conditions), and 13 isopathy. 133 clinical trials included a placebo control and 118 of those met all of our preset inclusion criteria, In 73% of these trials there was at least a trend in favour of homoeopathy, in 24% there was no difference, and in 3% placebo was better.

89 trials contained sufficient data for our preliminary meta-analysis. Overall, in the trials included, homoeopathy was statistically significantly superior to placebo. The results were similar when classical homoeopathy, clinical homoeopathy, isopathy, complex preparations and high potencies were analysed separately. When we analysed the results of those studies which met our predefined criteria for "better" quality, the superiority of homoeopathy versus placebo was less marked but still significant for most subgroups.

Discussion

Please, be aware that this is not truth, this is just a systematic review. Our results might have been distorted by publication bias (failure to publish undesired negative results), by flaws in the trials as well as in our review. Furthermore, we found no independently replicated trials using the same remedy, in similar patients and measuring similar outcomes.

However, we tried to control for biases as much as possible, and performed several sensitivity analyses to check the robustness of our results. In conclusion, the trials available are not compatible with the hypothesis that homoeopathy is a placebo.

But is the unbelievable, i.e. homoeopathy, now proven? And, if not, do you think that if ten more positive trials were published it then would be proved?

The Auto-Ganzfeld Debate

Maybe the socio-scientific processes and complexity of proving controversial hypotheses will become a little clearer if we look on other controversial areas of research.

Experiments in the area of extra sensory perception (ESP), in the sphere of para-psychology, deal with things that are even more unbelievable than homoeopathy. Their trials have to be even more rigorous, and looking at the literature shows how intriguing their studies are. I want to show you that even with very good empirical evidence it is hard to convince anyone.

Maybe the most straightforward experimental evaluations of ESP are the Auto-Ganzfeld experiments. In a very simplified manner they can be described as follows: The human subject on the receiving end of this experiment is sitting in an acoustically isolated room wearing headphones through which he hears white noise. He is perfectly relaxed and his eyes are covered. In another room a distance away, is the sender, a person like you and me, with a computer which selects a picture at random, from a stock of two hundred. The sender is asked to concentrate and to "transmit" the picture to the receiver for thirty minutes. At the end of 30 minutes the receiver is asked to identify the picture received which fits best his perceptions and feelings during the "transmission" phase from a set of four (the 3 "wrong" pictures again chosen by the computer at random). By chance, you would expect in a large series of such experiments that the "transmitted" picture is correctly identified 25% of the time. However, in several series of well-controlled studies up to 1985, the hit rate was about 35% and the statistical significance of the effect was "astronomic" (Honorton 1985).

Was this the proof that ESP exists?

Obviously - and absolutely correctly according to my opinion the results of these Ganzfeld experiments did not lead to an uncritical acceptance of ESP (Hyman 1985)] Science often has to face sensational claims supported by a certain body of evidence which are subsequently shown to be unjustified. And - also absolutely correctly in my opinion - you need more, and better evidence for "prowing" ESP or homoeopathy than for proving that aspirin is effective in decreasing headache. There is simply more background knowledge to make the effectiveness of aspirin plausible, than for effects of homoeopathic high potencies or for ESP.

What were the arguments of the sceptics?

The first argument was that the ESP results were not really positive, but mere fluctuations due to chance. This was due to the fact that originally the results were not presented in the easy and straightforward way above. However, when the analysis was restricted to the above-mentioned simple "hit rate" it became clear that this argument could not invalidate the findings.

An obvious second argument was the existence of publication bias. Although this argument could not be invalidated totally by the ESP-researchers, there were several points which made it unlikely. (1) the Parapsychological Association has had an official policy against selective reporting of "positive" results since 1975. I don't know if any other scientific society has yet come so far! (2) well-controlled Ganzfeld experiments need considerable financial and laboratory resources. Very few labs have the capacity to do such research in a reliable way; this fact - which is clearly a problem for independent replication - makes publication bias more unlikely. (3) the parapsychologists calculated the number of experimental sessions needed to nullify the findings by chance. Given the high degree of significance of the results, this number was so high that it was simply not reasonable to assume publication bias as the cause of the observed results.

Finally, sceptics suggested there were flaws in the experiments: sensory leakage, deception, fraud and so on. Compared with most clinical trials in conventional medicine or homoeopathy, the early Ganzfeld experiments were much more rigorous and straightforward. But as the results seemed so implausible, it was assumed that there simply had to be bias.

At this point something quite unique happened: The most prominent sceptics and proponents came together to make a joint communique on how future experiments should be done to be really convincing (Hyman 1986). In 1994, the results of these experiments were published in a major psychological journal. Still the results were "astronomically significant with a hit rate of about 35% (Bem and Honorton, 1994). Again the discussion started. Again statistics and methodology were discussed but there was little reason for serious doubt. Finally, the argument was put forward that the assumption that ESP exists is absurd and incompatible with scientific knowledge, and therefore, the results could not be true.

Although such reasoning is not scientifically reasonable it is quite common. The lesson to be learned from the Auto-Ganzfeld debate for homoeopathy is that one should not expect that by accumulating more and better empirical evidence, the controversy will be automatically solved.

If homoeopathy is a placebo . . .

Let us return to homoeopathy. The existing evidence seems to indicate that homoeopathy cannot be seen as a placebo in general. However, I want to raise the hypothesis: homoeopathy is a negligible. In the end, it might be that the effect of the remedy itself is not, or not very often, the major cause of improvements or changes.

Conclusion

We have seen that the available controlled clinical trials seem to provide evidence that homoeopathy cannot be regarded only as placebo. The example of ESP research shows that by simply accumulating good evidence it is difficult to solve controversies concerning "unbelievable" issues.

Finally, I tried to explain why the limitations in investigating the "specific" effects of a therapy might lead to a rejection of beneficial therapies.

Clinical research on homoeopathy has been focussed up to now mainly on the placebo question. In recent years a number of excellent trials of this type have been performed. Although most of these trials showed at least a tendency in favour of homoeopathy, the superiority over placebo was often disappointingly small. In most areas of clinical research effects sizes become smaller, the more the quality of the trials improve. I predict this for clinical research on homoeopathy too, even if homoeopathic remedies should cause specific effects. Doubleblind placebo-controlled trials are useful, but the limitations of that approach might result in misleading conclusions about clinical effectiveness.

Only a few trials have compared homoeopathic treatment strategies to standard therapies and very few of these trials used a pragmatic approach to testing homoeopathy in conditions similar to its actual practical use. There is a clear need for more pragmatic trials. However, there are a number of practical problems. (1) "Conventional" physicians are often reluctant to randomise their patients to a treatment which they consider to be a placebo. (2) Pragmatic trials are more open to bias, and given the controversial character of homoeopathy, will be strongly criticised. (3) and most important, there is not enough information on the practical use (conditions, inten enbons) and results (response rates) of homoeopathic treatments, to systematically identify the most relevant areas for research and to develop conclusive protocols. Up to now; most study models were developed "placebo". The homoeopathic materia medica is based on observations which are, at least in part, of doubtful validity Recent well controlled provings (Walach 1992) provide convincing evidence that volunteers receiving a placebo develop as many and very similar symptoms as those getting the remedies. I have not seen a rigorous proving which has shown convincingly that a remedy causes more or different symptoms compared to placebo. This does not mean that provings have not provided useful information, bot even for a critical proponent of homoeopathy - which I would consider myself to be - it is hard to imagine that all information in materia medicae and repertories is reliable. Another critical point is the impression that different homoeopaths seem to treat patients very differently, even in the same country. Are there really so many correct "similes" for a patient?

So, even if homoeopathy might work in principle, I hypothesise that in everyday practice it is often "only" a placebo. But I speculate that - at least in the case of good standard homoeopathy, taking seriously the patient as an individual - it can be effective.

How can this be?

Imagine a double blind controlled trial of an analgesic versus placebo for chronic pain. The analgesic might have improved the complaints in 50% of the patients, the placebo in 30%. In consequence the analgesic is 20% better than placebo. In a second hypothetical trial, a classical homoeopathic strategy is compared to placebo. In both groups 60% of patients might have improved; in consequence, it is no more effective than placebo. Imagine now that the results do not come from two separate but from one trial: Patients would be randomised to the two double-blind studies. The "homoeopathic placebo" would be more effective than the analgesic but "only a placebo". As a patient, which treatment would you prefer?

These trial results are pure speculation, but there is good evidence that response rates in placebo groups of similar trials vary greatly and have a relevant impact on results (see Moerman 1983). Individualised homoeopathic treatment involves a lot more than prescribing a pill and it might well be that a number of these processes (e.g. repeated intense anamnesis, instructions regarding nutrition) cause specific and beneficial effects. All of these are classified as "placebo" or "unspecific" effects, once the investigation begins to look at potentisation. All those who have experienced the atmosphere of a well done homoeopathic consultation, realise this possibility must not be regarded as on the basis of practicability and acceptability to methodologists. Furthermore, homoeopaths, overestimating at least the specific effects of their method, agreed to study models which were removed from their actual practice.

In conclusion, I think what is needed most at the moment is more descriptive information on the actual use of homoeopathy, on structures, processes and outcomes whether homoeopathy is "scientifically proven" or not, it is widely used. Homoeopaths should learn the basics of methodology, to learn that double-blind placebo-controlled trials are not the only tools available, and that the placebo question is not the only relevant one. Various questions are relevant, and to answer these a number of different and valuable tools exist and should be developed.

Research into homoeopathy should not only be seen as a cumbersome way to prove the unbelievable, but to describe, understand and improve actual performance.

Annotation

A more detailed presentation of this topic is given in (Linde 1997).

References

Bem DJ, Honorton C. Does Psi exist: Replicable evidence for an anomalous process of information transfer. Psycho Bull 1994; 115:4-18

Götzsche P. Trials of homoeopathy. Lancet 1993; 341:1533

Hopff WH Homöopathie kritisch betrachtet. Stuttgart: Thieme, 1991

Honorton C. Meta-analysis of Psi Ganzfeld research: a response to Hyman. J Parapsychol 1985; 49: 5-91

Hyman R. The Ganzfeld Psi experiment a critical appraisal. J Parapsychol 1985; 49: 3 49

Hyman R, Honorton C. A joint communique: The Psi Ganzfeld controversy. J Parapsychol 1986; 50:351-364

Kleijnen J, Knipschild P, ter Riet G. Clinical trials of homoeopathy. BMJ 1991; 302:316-323

Linde K, Clausius N, Ramirez G, Melchart D, Eitel F, Hedges LV, Jonas WB. Are the clinical effects of homoeopathy placebo effects? A meta-analysis of placebo-controlled trials. The Lancet, 1997;350:834-843.

Moerman D. General Medical effectiveness and human biology: placebo effects in the treatment. Med Anthropol Quat 1983;14:13-16

Müller F. Klinische und experimentelle Studien aus homöopathischer Medizin. Wien: Österreichische Gesellschaft für Homöopathische Medizin, 1991

Reilly DT, Taylor MAS, McSharry C, Aitchison T. Is homoeopathy a placebo response? Lancet 1986;ii:881-885

Walach H. Wissenschaftliche homöopathische Arzeimittelprüfung. Heidelberg: Haug, 1992

GLOSSARY

action	A dynamical quantity characterizing the activeness of a system. It is measured in unit of *Planck' constant.* Systems with high *action* may be treated according the laws of *classical physics.*
Avogadro's number	The number of molecules contained in one mole of a substance, i.e. N_A=6.022×10^{23} molecules/mole. If the concentration of a substance in a dilution is less than $(N_A)^{-1}$, then there is no molecule of the substance in the dilution.
Bose -Einstein condensation	If particles give up their classical distinguishability and individuality, they cross the border between the classical and the *quantum* world. This process is called Bose-Einstein condensation.
bosons	Particles whose statistics require them to be in states corresponding to symmetric *wave functions.* A fundamental property of particles following the Bose statistics is that they are indistinguishable. Examples are *photons* and *phonons.* Electrons are not bosons for they can be distinguished by their spin, electrons are *fermions.*
chaos	If we observe a physical system at some point of time and space and follow its evolution in time, ie. its dynamics, a second observation of the physical system starting at the same point with only the tiniest deviation from the initial starting point, would usually lead to a very similar observation of the dynamics of the system. The physical system is then called deterministic. A physical system is chaotic, if small errors in the initial conditions are amplified very rapidly, leading to apparently random behaviour, unrelated in any simple way to what we imagined were the starting point of our observation.
classical physics	Strictly deterministic (opposed to e.g. quantum uncertainty) physics constructed according to laws of Newton and Maxwell. It provides a good description of the behavior of systems whose *action* is high compared with *Planck's constant.*

coherence — When two or more waves or oscillations are in perfect alignment with each other, they are said to be in phase. A group of waves or oscillations having all the same frequency and same phase is called coherent waves (coherent light) or coherent oscillation. Coherence is one of the key properties of laser light. The light waves emitted by a laser remains coherent (or in phase) over many inches distance before they become out of step due to small differences in their original wave length (negligible spread).

de Broglie wave — In order to preserve the relativistic symmetry in the formalism of special relativity where time and space are set on equal footing, de Broglie formulated a relation between momentum and wave length. The Planck formula $E = hf$ makes energy proportional to the frequency f, the number of vibrations per unit interval of time (h is the ***Planck's constant)***, a relation that can be derived in the space-time formulation of special relativity. The momentum-energy formulation of de Broglie's interpretation leads to the conclusion $p = h / \lambda$ where momentum p is proportional to the inverse of wave length λ, that is the number of vibrations per space interval. Here, it becomes obvious that the wave -particle dualism (particle properties E,p related to wave properties f,λ) does not means an "either" or "or". It actually shows how limited our classical imagination is to express the property of energy (or matter) in Newton and Maxwell images of *classical physics*. The de Broglie wave thus can be assigned to any quantum of mass or energy. The ration of the de Broglie wave length and the physical size of the object is a measure of the quantumness of the object.

entropy — A measure of a system's order or information contents. According the second law of thermodynamics, change of entropy is always positive (i.e. information content decreases), if energy is supplied.

fermions — Particles whose statistics require them to be in states corresponding to anti-symmetric *wave functions*. Examples are electrons, protons and neutrons.

mesoscopic — The mesoscopic range of the physical size scale is region where *quantum* phenomena are about to loose their prominence and the simpler rules of continuum physics are about to take over.

observable — A quantity associated with a physical system which can be measured experimentally. Examples are location, momentum, time and energy.

phonon — A *quantum* of vibrational energy. As it can not be observed as a particle-like object, it is sometimes called quasi-particle. A phonon belongs to the boson family of particles.

photon — A *quantum* of radiation energy (light). The energy of a photon is proportional to the frequency of its radiation. A photon belongs to the boson family of particles.

Planck's constant — The fundamental unit of *action* (h=6.63x10^{-34} Joule-seconds) which sets the scale for large and small in *quantum* theory.

quantum — Literally, the unit of a quantity, the smallest definable amount by which to measure a given physical system. It stands for something that is discrete and countable. Cents are the quanta of money. Each grain of sand is a quantum of what otherwise appears to be a continuous expanse of beautiful Australian beach. In physics, quantum stands for an elemental unit of energy, a "grain" of an energy of atoms, molecules and electromagnetic radiation (light).

quantum electrodynamics — The theory of the interaction of *photons* and *electrons*. Quantum electrodynamics (QED) is the most fundamental and accurate theory in modern physics.

Schrödinger equation — The fundamental equation of motion in quantum *wave mechanics* governing the evolution of a quantum physical system.

spin — The intrinsic angular momentum carried by all *quantum* particles like e.g. electrons and *photons*. Electrons, like all *fermions*, have half integer spin, photons, like all *bosons*, have integral integer spin.

state — Characterizes motion of a system. A state of a system is characterized by as a much as possible details of its motion.

wave function — Represents the *state* of a physical system in quantum theory (*wave mechanics* or quantum mechanics). It is a mathematical abstract representing the dynamics of the particles of the quantum world, eg. *photons* and *phonons*.

wave mechanics — The version of quantum theory, introduced by Schrödinger, which emphasizes the wave-like character of physical systems. It represents the purely mathematical descriptions of the dynamics found in the *quantum* world.

wave-particle duality — Free electrons and *photons* behave both as they were particles and also as if they were waves. Many text books still explain that the nature of electrons and photons is determined by the experiment being set up for their measurement. This unfortunate simplification has caused many misconceptions about wave or particle nature of energy or matter, especially among non-physicists.

homoeopathy Medical treatment of or experimental influence on a living system according to the *Law of Similarity* . Also a common expression for the use of specially prepared dilutions (*preparation*) in medical therapy or in scientific experiments.

initial aggravation Cybernetic reaction of the organism after application of a homoeopathic dilution.

individualisation Individualized choice of a homoeopathic remedy with regard to the human patient's peculiar symptoms, with which not only the pathological clinical symptoms, but also further characteristics are meant. In research on lower animals or plants, the 'individual' reaction of different species is discussed.

law of similarity Both the application of a molecular (crude) substance or a homoeopathic dilution of that substance on healthy persons (*remedy proving*) as well as an independent pathological process can induce similar, clearly defined symptoms. The very homoeopathic dilution that is able to induce the defined symptoms in healthy persons (*remedy proving*) is used in medical therapy of the analogous symptom (that has occurred due to the independent pathological process). Two kinds of examples are given:
1 Cure after exposition to or protection from a toxin or allergen by a homoeopathic preparation of this toxin or allergen.
2 Both a bee sting as well as ultra-violet rays can induce an inflammatory reaction. Homoeopathic dilutions of bee venom are inhibit cutaneous erythema induced both by bee venom as well as by ultraviolet rays.

potency rule Through the *preparation* process, information is transferred to the solvent. It is assumed that this information is more marked, the more often the process of agitation and dilution has taken place.

preparation potentisation potenziation The Process of stepwise dilution of a substance e.g. in water or a water-alcohol mixture and input of exogenic energy by agitation succussing or vortexing) between the dilution steps. Non-watersoluble substances are first triturated with lactose. D- dilutions are diluted in steps of 1:10, C-dilutions in steps of 1:100.

remedy proving Drug Prooving. Application of a crude (molecular) substance or a high or *ultra high dilution* to healthy volunteers with the aim to provoke symptoms (empirical homoeopathic intoxication studies).

sensitization	Homoeopathic dilutions, like other (weak) stimuli, obviously exert their effects only in biological systems which are sensitive due to pathological, biological or experimental stress conditions.
succussion	The vigorous shaking of the solution during the *potentization* process of a *homoeopathy* remedy is known as succussion.
ultra high dilution UHD	Homoeopathically prepared dilution of infinitesimal small quantities of the original (crude) substance (sometimes simply called 'high dilutions'), in general diluted a concentration of less than one original substance molecule per mole. In ultra high dilutions characteristic properties of the original substance are carried over from dilution step to dilution step, instead of substance molecules themselves which is physically impossible at such low concentration because of Avogadro's number.

LIST OF AUTHORS

Prof Bastide Madeleine
Department of Pharmacy (Immunology)
University of Montpellier
France

Prof Paolo Bellavite
Istituto di Chimica e
di Microscopia Clinica
Ospedale Policlinico
37134 Verona
Italy

Dr Massimo Citro
Istituto di Ricera Metamolecular
Via Cibrario 33
10143 Torino
Italy

Prof Emilio Del Giudice
Department of Nuclear Physics INFN
University of Milano
Via Celoria 16
20133 Milano
Italy

Dr P Christian Endler
Ludwig-Boltzmann Research Laboratory
for Low Energy Biophysics
8010 Graz
Austria

Dr Peter Fisher
Royal London Homoeopathic Hospital
The Faculty of Homoeopathy

Dr Mae-Wan Ho
Open University Walton Hall
Bioelectrodynamics Laboratory
Milton Keynes, MK76AA
Great Britain

Dr Egon Laupert
Ludwig-Boltzmann Research
Laboratory
for Low Energy Biophysics
8010 Graz
Austria

Dr Klaus Linde
University of Munich
Munich
Germany

Dr Andrea Nograsek
Institute for Botanics
University of Graz
8010 Graz
Austria

Dr Waltraud Pongratz-Scherer
Ludwig-Boltzmann Research
Laboratory
for Low Energy Biophysics
8010 Graz
Austria

Prof G Preparata
Department of Nuclear Physics INFN
University of Milano
Via Celoria 16
20133 Milano
Italy

Dr Jurgen Schulte
University of Technology, Sydney
Department of Applied Physics
POB 123, Broadway 1
Sydney, NSW 2007
Australia
and
Ludwig-Boltzmann Research
Laboratory for Low Energy Biophysics
8010 Graz
Austria

Dr Cyril W Smith
Department of Electric and
Electronic Engineering
University of Salford
827221 Salford
Great Britain

Dr Roeland Van Wijk
Department of Molecular Cell Biology
University of Utrecht
Padualaan 8
3584 CH Utrecht
The Netherlands

Dr Fred AC Wiegant
Van Wijk Roeland
Department of Molecular Cell Biology
University of Utrecht
Padualaan 8
3584 CH Utrecht
The Netherlands

INDEX

—I—

—L—

—M—

—N—

—O—

—P—

—Q—

—R—

—S—

—T—

—U—

—V—

—W—

MIX
Papier aus verantwortungsvollen Quellen
Paper from responsible sources
FSC® C105338

If you have any concerns about our products,
you can contact us on
ProductSafety@springernature.com

In case Publisher is established outside the EU,
the EU authorized representative is:
Springer Nature Customer Service Center GmbH
Europaplatz 3, 69115 Heidelberg, Germany

Printed by Libri Plureos GmbH
in Hamburg, Germany